Springer Tracts in Advanced Robotics

Volume 3

Springer

Berlin
Heidelberg
New York
Hong Kong
London
Milan
Paris
Tokyo

Ciro Natale

Interaction Control of Robot Manipulators

Six degrees-of-freedom tasks

With 36 Figures

Springer

Professor Bruno Siciliano, Dipartimento di Informatica e Sistemistica, Università degli Studi di Napoli Federico II, Via Claudio 21, 80125 Napoli, Italy, email: siciliano@unina.it

Professor Oussama Khatib, Robotics Laboratory, Department of Computer Science, Stanford University, Stanford, CA 94305-9010, USA, email: khatib@cs.stanford.edu

Professor Frans Groen, Department of Computer Science, Universiteit van Amsterdam, Kruislaan 403, 1098 SJ Amsterdam, The Netherlands, email: groen@science.uva.nl

STAR (Springer Tracts in Advanced Robotics) has been promoted under the auspices of EURON (European Robotics Research Network)

Author

Dr. Ciro Natale
Seconda Università degli Studi di Napoli
Dipartimento di Ingegneria dell'Informazione
Via Roma 29
81031 Aversa (CE), Italy

ISSN 1610-7438

ISBN 3-540-00159-X Springer-Verlag Berlin Heidelberg New York

Cataloging-in-Publication Data applied for
Bibliographic information published by Die Deutsche Bibliothek
Die Deutsche Bibliothek lists this publication in the Deutsche Nationalbibliografie; detailed bibliographic data is available in the Internet at <http://dnb.ddb.de>.

Springer-Verlag Berlin Heidelberg New York
a member of BertelsmannSpringer Science + Business Media GmbH

http://www.springer.de

Typesetting: Digital data supplied by author. Data-conversion by PTP-Berlin e.k., Stefan Sossna
Cover-Design: design & production GmbH, Heidelberg
Printed on acid-free paper 62/3020 Rw - 5 4 3 2 1 0

Foreword

The field of robotics continues to flourish and develop. In common with general scientific investigation, new ideas and implementations emerge quite spontaneously and these are discussed, used, discarded or subsumed at conferences, in the reference journals, as well as through the Internet. After a little more maturity has been acquired by the new concepts, then archival publication as a scientific or engineering monograph may occur.

The goal of the Springer Tracts in Advanced Robotics is to publish new developments and advances in the fields of robotics research – rapidly and informally but with a high quality. It is hoped that prospective authors will welcome the opportunity to publish a structured presentation of some of the emerging robotics methodologies and technologies.

The monograph written by Ciro Natale is focused on a central problem in robot control, namely that to manage the interaction between a robot manipulator and the environment. Indeed, providing the control system with both motion and force control capabilities becomes crucial for successful execution of complex tasks. Its foreseen impact on the performance of the next generation of robots is starting to be recognized by major manufacturers world-wide. To this regard, the unique feature of the work lies in its comprehensive treatment of the problem from the theoretical development of the various schemes down to the real-time implementation of interaction control algorithms on industrial robots, with extensions even to a dual robot cooperative system.

Remarkably, the doctoral thesis at the basis of this monograph was a finalist for the First EURON Georges Giralt PhD Award devoted to the best PhD thesis in Robotics in Europe. A fine addition to the series!

Napoli, September 2002 *Bruno Siciliano*

Preface

My first contact with the 'force' dates back to 1977, when Obi-Wan Kenobi amazed me with the famous motto "may the force be with you"... I looked in astonishment at the screen and since then my future has been guided by the power of force!

This monograph gathers all the research work carried out for my PhD thesis, supervised by Prof. Bruno Siciliano, which was selected among the finalists of the *First EURON PhD Award* last year. It is the result of three fruitful years spent in collaboration with the robotics group of the PRISMA Lab at the University of Naples Federico II.

The problem of robot force control has been investigated in depth by many researchers for more than twenty years. However, very few monographs exist treating the problem of interaction tasks involving both contact forces and contact moments unless some simplifying conditions are assumed. On the other hand, controlling the interaction forces and moments during real industrial applications, such as deburring, assembling, polishing, could enhance the result of the task in terms of accuracy and cycle time. But this requires control algorithms efficient and effective in a broad variety of working conditions.

Therefore, this book is aimed at developing a treatment of the interaction control problem in the general case of a six degrees-of-freedom task foregoing the hypothesis of small displacements. Moreover, a breakthrough of the work lies on the steps towards the control of cooperative robots interacting with each other and with the working environment. Many applications can draw advantage from the coordination of multiple robots performing the same task, e.g. carrying large or heavy loads, parts mating, assembling. Even though this is still an emerging topic, this book can be considered a first attempt at developing a methodology framed in the task space approach, as a strategy to deal with the problem of interaction of both a single manipulator and multiple robots with the environment. The problem of multiple robot coordination is tackled both at task planning level and at feedback control level, by resorting to a modular control architecture, very attractive from an industrial point of view.

The contents of the monograph are organized as follows. Chapter 1 presents a classification of robotic interaction tasks and evidences the key points of the task space approach; a brief description of the laboratory setup used to experiment all the proposed control algorithms is reported at the end of the chapter. Chapter 2 is devoted to presenting some motion control schemes in the task space, each one corresponding to a different representation of end-effector orientation; they are experimentally compared in terms of performance and computational complexity and constitute the first stage of an interaction controller. A strategy for indirect force control is presented in Chapter 3, where the concept of six degrees-of-freedom impedance is introduced on energy-based arguments and with reference to the particular representation of end-effector orientation. The impedance control scheme is then designed and experimental results are reported. The direct force control strategy is developed in Chapter 4, where a parallel approach is pursued to handle the interaction with a scarcely structured environment. The last Chapter is devoted to tackling the problem of controlling a dual-arm robotic system interacting with a compliant environment, and different strategies are presented to deal with the coordination problem. The first Appendix provides the necessary background on rigid body orientation and reports the definition and the main properties of the unit quaternion. The second Appendix reports some notes about real-time implementation of interaction control algorithms on industrial robots.

The book is mainly addressed to researchers and scholars who have their first contact with robot force control. At the same time, it can be adopted as a textbook for a graduate course on advanced robotics requiring, as background knowledge the basics of kinematics and dynamics of robot manipulators, as well as feedback control theory.

Aversa, September 2002 *Ciro Natale*

Acknowledgements

I wish to dedicate my first acknowledgement to the advisor of my PhD thesis, Prof. Bruno Siciliano, for having introduced me to the amazing world of scientific research in robotics and for having encouraged me to write "my first book".

Particular thanks go to Prof. Lorenzo Sciavicco. Without his invaluable effort in the establishment of PRISMA Lab at the University of Naples Federico II, not even a line of this book would have ever been written.

Prof. Luigi Villani and Prof. Fabrizio Caccavale deserve a special mention as co-authors of many scientific articles where the results of this work have been published.

I am very grateful to Prof. Gerd Hirzinger and all the colleagues of the German Aerospace Center, in particular Dr. Ralf Koeppe, for having hosted me in one of the most exciting and challenging robotics laboratories I have ever visited, and for the appreciation of my work they demonstrated so many times.

Of course, "at this time" I can feel grateful to the Commission and the anonymous reviewers of the First Euron PhD Award for having expressed a so gratifying judgement of this work.

Finally, a word of appreciation is for Ms. Penelope Thresh for her annotations on a draft of this book.

Aversa, September 2002 *Ciro Natale*

Contents

1. Introduction

In this introductory chapter, the problem of controlling a robot manipulator which performs six-degrees-of-freedom (six-DOF) tasks requiring interaction with the environment is described. The difference between the well-assessed operational space approach and the task space approach pursued here is explained. A detailed classification of different control problems in the task space framework is carried out and, finally, a brief description of the experimental setup adopted to validate all the algorithms presented through the book is reported.

1.1 Six-DOF Interaction Tasks

Control of interaction between a robot manipulator and the environment is crucial for successful execution of a number of practical tasks where the robot end effector has to manipulate an object or perform some operation on a surface. Typical examples include polishing, deburring, machining or assembling. A complete classification of possible robot tasks is practically infeasible in view of the large variety of cases that may occur, nor would such a classification be really useful to find a general strategy to control *interaction with the environment*.

On the other hand, a certain classification can be based on the number of degrees of freedom involved. During interaction, the environment sets constraints on the geometric paths that can be followed by the end effector. This situation is generally referred to as *constrained motion*. When only the translational degrees of freedom of the motion are constrained, the interaction task can be classified as a *three-DOF task* because only *linear forces* may arise during its execution. On the other hand, if the motion is constrained along both the translational and the rotational degrees of freedom, the interaction task can be classified as a *six-DOF task*, and both forces and *torques* may arise during the task execution. In the remainder, the term torque will be often referred to simply as *moment*, since a torque is the moment of linear force with respect to a pole.

In order to describe and successfully perform six-DOF interaction tasks, a suitable description of end-effector orientation should be adopted. Two main different approaches can be distinguished, based on the chosen type of representation.

1.1.1 Operational space approach

Since the number of variables necessary to describe a six-DOF interaction task is six, the most widely used approach in literature to represent the rotational degrees of freedom is to adopt a minimal representation of end-effector orientation, i.e. a set of three *Euler angles*. In fact, in this way a vector of six components comprising the three Euler angles and the three position variables is defined to describe the six-DOF *pose* of the end effector during task execution. Such a technique is usually referred to as the *operational space approach* [45].

Two main drawbacks can be ascribed to such a formulation, namely the occurrence of *representation singularities* and inconsistency with the *task geometry.* The former is due to a topological problem concerning the parametrization of the set of rotations (see Appendix A). The latter can be imputed to the fact that the set of three Euler angles does not represent a vector in the Cartesian space, so that it is not possible to let an orientation displacement correspond to a moment vector in a geometrically consistent fashion. In other words, if a task requires a moment to be applied along a certain axis in the Cartesian space, the corresponding orientation displacement in terms of three Euler angles is not always aligned to the same axis, apart from some special cases, e.g., when the principal axes of the end-effector frame are aligned with the axes of the base frame and the actual orientation is far from a representation singularity. On the other hand, the property according to which a formulation allows the moment applied during the task execution to be aligned with the induced orientation displacement is here termed *task geometric consistency*.

1.1.2 Task space approach

In order to overcome the drawbacks of the approach described in the previous section it is worth resorting to a geometric meaningful representation of end-effector orientation. The only descriptions for which the property of task geometric consistency holds, are those adopting the so-called angle/axis approach [16], and the corresponding formulation describing the interaction task is here called *task space approach.*

The first characteristic of the above representations of orientation is that they have a geometrically meaningful interpretation, and thus the property of task geometric consistency holds. The second main feature is that special non-minimal representations, like the *unit quaternion*, belong to this class of orientation descriptions; hence, the problem of representation singularities does not exist (see Appendix A).

The first step to be performed for executing a robotic manipulation task is to assign a desired trajectory to be tracked by the robot end effector. For this purpose, the properties of the task space approach will be usefully utilized to perform motion control of the robot manipulator, so as to fully exploit the geometric consistency property for the specification of the desired trajectory.

1.2 Task Space Motion Control

The techniques to perform motion control in the task space will be examined. A comparison among the well-known control schemes will be carried out in order to understand the main features of each approach and identify the characteristics more relevant for the purpose of interaction control as will be illustrated further.

1.2.1 Resolved acceleration control

The most widely adopted control technique in the literature to design a motion control for a generic mechanical system is based on the inverse dynamics strategy, thanks to the peculiar properties of the dynamic model of a mechanical system [63]. The main consequence of such a technique is that the control problem can be translated into the problem of designing a suitable acceleration profile for the mechanical system —in this work the robot end effector— so as to track the desired trajectory, while guaranteeing stability and robustness properties.

In the operational space approach, the acceleration to be designed is the second time derivative of the six-dimensional vector expressing the end-effector pose, which does not have any physical meaning as regards the rotational part. Instead, in the task space approach, the acceleration to be designed is the acceleration vector constituted by the *linear* and the *angular* accelerations. This allows the user to specify a trajectory in terms of a translation along and a rotation about the axes in the Cartesian space, with a clear geometric meaning.

1.2.2 Representation singularities

Apart from the geometric meaning of the desired trajectory to be tracked, another problem may arise by using operational space approaches based on minimal representations of orientation, i.e., the occurrence of representation singularities. This can be overcome by adopting non-minimal representations such as the unit quaternion (see Appendix A). From a control point of view, this means that the control laws can be always well-defined and the closed loop system can be globally stable.

1.3 Task Space Interaction Control

A purely motion control strategy for controlling interaction is a candidate to fail executing robotic interaction tasks. In fact, successful execution of an interaction task with the environment by using motion control could be only obtained if the task were accurately planned. This would in turn require an accurate model of both the robot manipulator (kinematics and dynamics) and the environment (geometry and mechanical features). Manipulator modelling can be known with enough precision, but a detailed description of the environment is difficult to obtain.

To understand the importance of task planning accuracy, it is sufficient to observe that to perform a mechanical parts mating with a positional approach, the relative positioning of the parts should be guaranteed with an accuracy of an order of magnitude greater than part mechanical tolerance. Once the absolute position of one part is exactly known, the manipulator should guide the motion of the other with the same accuracy.

In practice, the planning errors may give rise to a *contact force* causing a deviation of the end effector from the desired trajectory. On the other hand, the control system reacts to reduce such deviation. This ultimately leads to a build-up of the contact force until saturation of the joint actuators is reached or breakage of the parts in contact occurs.

The higher the environment stiffness and position control accuracy are, the easier a situation like the one just described can occur. This drawback can be overcome if a *compliant behaviour* is ensured during the interaction. This can be achieved either in a *passive* fashion by interposing a suitable compliant mechanical device between the manipulator end effector and the environment, or in an *active* fashion by devising a suitable *interaction control* strategy.

The *contact force* is the quantity describing the state of interaction in the most complete fashion. Therefore, it is expected that enhanced performance can be achieved with an interaction control provided that force measurements are available. To this purpose, a *force/torque sensor* can be mounted on a robot manipulator, typically between the wrist and the end effector, and its readings shall be passed to the robot control unit via a suitable interface.

Robot force control has attracted a wide number of researchers in the past two decades. A state-of-the-art of the first decade is provided in [71], whereas the progress of the last decade is surveyed in [68] and [29]. Very recently, a monograph on force control [37] has appeared, but only a second one deals with six-DOF tasks performed by a single robotic arm [65].

Interaction control strategies can be grouped in two categories; those performing *indirect force control* and those performing *direct force control*. The main difference between the two categories is that the former achieve force control via motion control, without explicit closure of a force feedback loop; the latter, instead, allows controlling the contact force to a desired value, owing to the closure of a force feedback loop.

1.3.1 Impedance control

Impedance control [39], belonging to the first category, is conceived so as to relate the position error to the contact force through a mechanical impedance of adjustable parameters. A robot manipulator under impedance control is described by an equivalent mass-spring-damper system with the contact force as input. The resulting impedance in the various task space directions is typically nonlinear and coupled. If a force/torque sensor is available, then force measurements can be used in the control law so as to achieve a linear and decoupled impedance in the translational directions. Concerning the rotational directions, the definition of the impedance equation depends on the chosen formulation, namely, the operational space approach or the task space approach. In this work, it will be demonstrated how the former leads to task geometric inconsistency, and thus the latter will be formalized and its performance will be analysed from both a theoretical and an experimental point of view. Interestingly enough, the proposed rotational impedance equation will result nonlinear, however it will preserve task geometric consistency.

The main limitation of impedance control is that it is not possible to regulate the contact force to a desired value; such a limit is overcome by control techniques belonging to the second category, as described below.

1.3.2 Force control

If a detailed model of the environment is available, a widely adopted strategy of force control is the *hybrid position/force control* which aims at controlling position along the unconstrained task directions and force along the constrained task directions. A selection matrix acting on both desired and feedback quantities serves this purpose for typically planar contact surfaces [61], whereas the explicit constraint equations have to be taken into account for general curved contact surfaces [72, 52, 53].

In most practical situations, a detailed model of the environment is not available. In such a case, an effective strategy still in the second category is the *inner/outer motion/force control* where an outer force control loop is closed around the inner motion control loop which is typically available in a robot manipulator [30]. In order to embed the possibility of controlling motion along the unconstrained task directions, the desired end-effector motion can be input to the inner loop of an inner/outer motion/force control scheme. The resulting *parallel control* is composed of a force control action and a motion control action, where the former is designed so as to dominate the latter in order to ensure force control along the constrained task directions [25].

It should be clear that the inner/outer approach can be pursued only in case of contact with a more or less compliant environment. In fact, the interaction with a rigid surface constrains the motion of the end effector by reducing the number of its degrees of freedom, and thus the contact forces are reactions of the constraint surface and they can be controlled only by directly acting on the torques of the actuators. On the other hand, in the interaction with a compliant surface, the contact force can be related to end-effector displacements, and thus it can be controlled by controlling end-effector motion also along the constrained directions.

In this work, only interaction with compliant environments is of interest, for two main reasons. First, torque control is effective only if a special robot is available or the robot is equipped with torque sensors, as shown in the literature [5], and this is not the case of the common industrial robots as those available in the PRISMA Lab. Moreover, to manage the interaction with rigid surfaces many solutions can be devised; one for all is the adoption of a passive compliance, for example placed directly in the force/torque sensor. The effectiveness of the approach has been demonstrated in some works related to integration of force control into industrial robots, e.g., [56, 43].

In order to validate the theoretical findings, every interaction control scheme presented throughout this book is tested in a number of experiments for representative interaction tasks. Indeed, the force/torque sensor needed for the execution of interaction control schemes is not typically available for industrial robots, because of high cost and scarce reliability. Commercially available industrial robots are purposefully utilized to demonstrate the credibility of force control in the perspective of the next generation of robot control units with enhanced sensory feedback capabilities. The problem of interfacing the force/torque sensor has been solved thanks to the open architecture of the control unit. This feature is crucial also for the implementation of control schemes requiring model-based compensation actions. Therefore, a more detailed description of the architecture is developed in the following section.

1.4 PRISMA Lab Experimental Setup

The *experimental setup* available in the PRISMA Lab comprises two industrial robots Comau SMART-3 S (Fig. 1.1), with two control units C3G 9000, equipped with two force/torque sensors ATI FT130/10.

1.4.1 Robot manipulators

Each robot manipulator has a six-revolute-joint anthropomorphic geometry with nonnull shoulder and elbow offsets and non-spherical wrist. One manipulator is mounted on a sliding track which provides an additional degree of mobility. The joints are actuated by brushless motors via gear trains; shaft absolute resolvers provide motor position measurements, and no tachometers are available.

Each robot is controlled by the C3G 9000 control unit which has a VME-based architecture with 2 processing boards (Robot CPU and Servo CPU) both based on a Motorola 68020/68882, where the latter has an additional

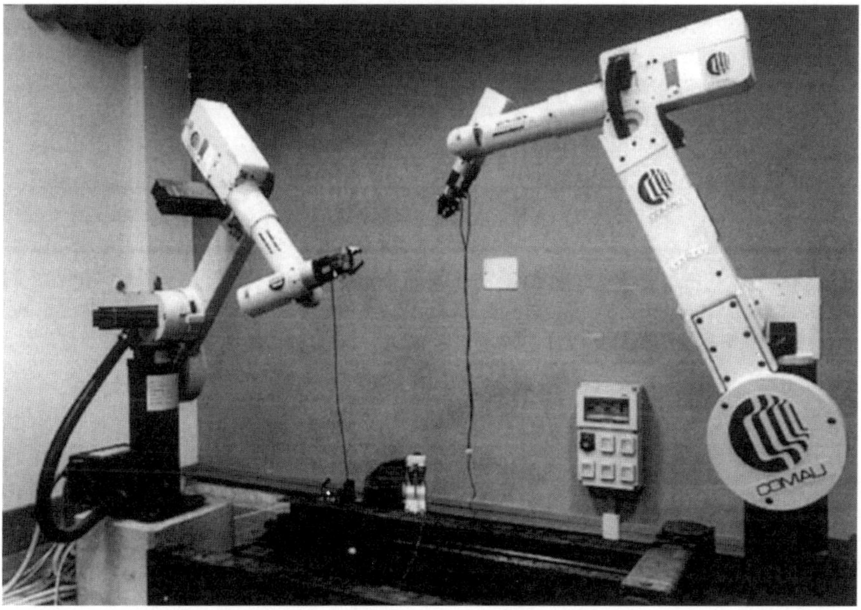

Fig. 1.1. Experimental setup available in the PRISMA Lab

DSP and is in charge of trajectory generation, inverse kinematics and joint position servo control. Independent joint control is adopted where the individual servos are implemented as standard PID controllers. The native robot programming language is PDL 2, a high-level Pascal-like language with typical motion planning instructions [66].

1.4.2 Open control architecture

An *open control architecture* is available which allows testing of advanced model-based control algorithms on a conventional industrial robot [31]. Connection of the VME bus of the C3G 9000 unit to the ISA bus of a standard PC is made possible by a BIT 3 Computer bus adapter board, and the PC and C3G controller communicate via the shared memory available in the Robot CPU; the experiments reported in the present book have been carried out by using a PC Pentium MMX/233. Time synchronization is implemented by interrupt signals from the C3G to the PC with data exchange at a given sampling rate. A set of C routines are available to drive the bus adapter boards. These routines are collected in a library (PCC3Link) and are devoted to performing communication tasks, e.g., reading shaft motor positions and/or writing motor reference currents from/to the shared memory. Also, a set of routines are devoted to performing safety checks and monitoring system health, e.g., a watchdog function and/or maximum current checks.

Seven different operating modes are available in the control unit, allowing the PC to interact with the original controller both at trajectory generation level and at joint control level. To implement model-based control schemes, the operating mode number 4 is used in which the PC is in charge of computing the control algorithm and passing the references to the current servos through the communication link at 1 ms sampling time. Joint velocities are reconstructed through numerical differentiation of joint position readings. All control algorithms are implemented in discrete time at 1 ms sampling interval, except when both robots are controlled in the open operating mode; in this case, a 2 ms sampling interval has to be set for synchronization reasons.

A schematic of the open control architecture is depicted in Fig. 1.2.

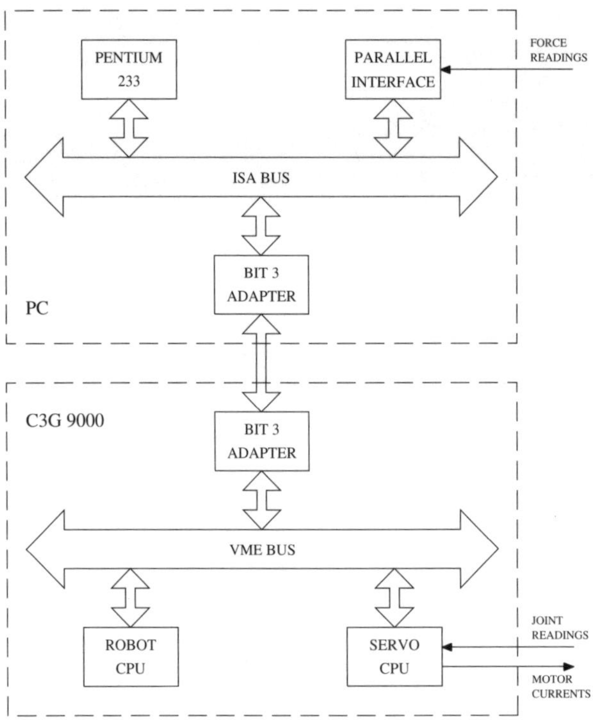

Fig. 1.2. Schematic of open control architecture

1.4.3 Force/torque sensor

A six-axis *force/torque sensor* ATI FT130-10 with force range of ± 130 N and torque range of ± 10 Nm can be mounted at the wrist of either robot (Fig. 1.3). The sensor is connected to the PC by a parallel interface board which provides

Table 1.1. Force/torque sensor technical data

Force range	±130 N
Torque range	±10 Nm
f_x, f_y resolution	0.1 N
f_z resolution	0.2 N
μ_x, μ_y, μ_z resolution	0.005 Nm
k_x, k_y stiffness	$8.8 \cdot 10^6$ N/m
k_z stiffness	$17 \cdot 10^6$ N/m
k_{tx}, k_{ty} stiffness	$10 \cdot 10^3$ Nm/rad
k_{tz} stiffness	$16 \cdot 10^3$ Nm/rad
f_x, f_y max load	650 N
f_z max load	1300 N
μ_x, μ_y, μ_z max load	50 Nm

readings of six components of generalized force at 1 ms, citeatift. The main technical data of the device are summarized in Tab. 1.1.

The type of environment considered in the various case studies presented in the following chapters consists of planar compliant surfaces of medium stiffness, with an estimated contact stiffness of the order of 10^4 N/m for translational displacements. As for the rotational displacements, contact moments are obtained by using special tools mounted to the end effector, like disks or hollow parts for assembly tasks. The resulting rotational stiffness is of the order of 10 Nm/rad. This choice is motivated by the desire of safely analysing the performance of the various control schemes where the interaction with the environment encompasses an unplanned transition from non-contact to contact at nonnegligible end-effector speed and taking into account that the sensor is highly stiff (see Tab. 1.1).

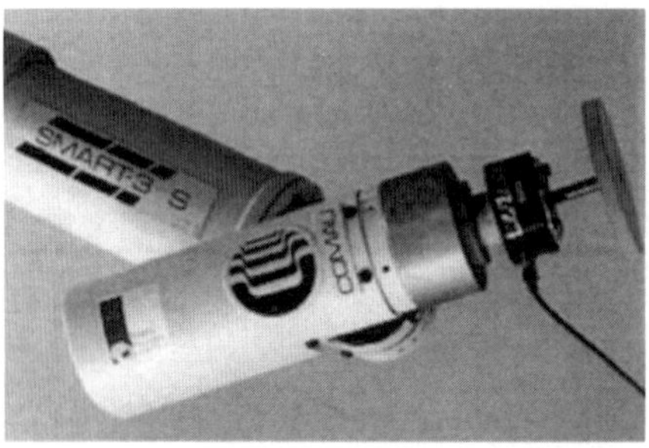

Fig. 1.3. Force/torque sensor mounted at robot wrist

More details about real-time implementation of the interaction control algorithms, which will be presented in the following chapters, and force sensor interfacing can be found in Appendix B.

The theoretical and experimental results presented in this work have been published in some international journals as well as in proceedings of several international conferences. The task space motion control problem has been reviewed in [14]. The task space impedance control was originally proposed in [13] and more generally formalized in [16]; the extension to redundant manipulators has been proposed in [58]. The task space parallel position/force control has been extended to six-DOF interaction tasks in [57]. The results of applications on the dual-robot system have been published in [15, 17] as concerns the loose cooperative control, and as regards the tight cooperative control they have been presented in [18, 19].

2. Task Space Motion Control

The first step towards interaction control of a robot manipulator is setting up a good dynamic model and design a robust and efficient motion controller. The first issue is discussed in the first section, where the notations used in the remainder of the book are also presented, while the problem of motion control design is addressed in the second section. The task space approach for motion control is based on the well-known resolved acceleration technique, but different algorithms are critically compared from different point of views, namely the adopted representation of end-effector orientation and the computational complexity. Finally, an experimental verification of the theoretical conclusions is carried out.

2.1 Modelling

In every control problem an accurate model of the system to be controlled is very helpful to the purpose of controller design. In this work, the system is constituted by one or two robot manipulators and a passive environment, that is a mechanical system is of concern. This greatly simplifies the modelling problem since the structure of the mathematical model is almost always known and very effective procedures can be adopted to accurately identify the parameters.

Two types of parameters can be distinguished in the model of a robot manipulator, the *kinematic parameters* and the *dynamic parameters*. The former are usually known quite accurately from the CAD model of the robot, while the latter have to be identified. For this purpose, many well-established techniques can be found in literature; the dynamic parameters of both robots available in the PRISMA Lab have been successfully identified by using the general systematic identification procedure proposed in [4].

A robot manipulator consists of a kinematic chain of $n+1$ links connected by means of n joints. One end of the chain is connected to the base link, whereas an end effector is connected to the other end.

Let $\boldsymbol{q}$ denote the $(n \times 1)$ vector of *joint variables*, $\Sigma_b(O_b–X_bY_bZ_b)$ and $\Sigma_e(O_e–X_eY_eZ_e)$ be a frame attached to the base of the robot—*the base frame*—and a frame attached to the end effector—*the end-effector frame*, respectively.

The *position* of the end-effector frame with respect to the base frame is the position of the origin O_e represented by the 3×1 vector $\boldsymbol{p}_e$, while the *orientation* of the end-effector frame with respect to the base frame can be represented in many ways as described in Appendix A, e.g. a 3×3 rotation matrix $\boldsymbol{R}_e = [\,\boldsymbol{n}_e \quad \boldsymbol{s}_e \quad \boldsymbol{a}_e\,]$ or a unit quaternion $\mathcal{Q}_e = \{\eta_e, \boldsymbol{\epsilon}_e\}$. Notice that no superscript has been used; in fact, hereafter, a superscript will precede the vector or the matrix quantity only if this will be referred to a frame other than the base frame. Regarding the unit quaternions, a first subscript will denote the frame whose orientation is represented by the quaternion itself and a second subscript will denote the reference frame, hence it will be dropped when referring to the base frame.

The *kinematic model* of the manipulator gives the relationship between $\boldsymbol{q}$ and $\boldsymbol{p}_e$, i.e.

$$\boldsymbol{p}_e = \boldsymbol{p}_e(\boldsymbol{q}) \tag{2.1}$$

as well as between $\boldsymbol{q}$ and $\boldsymbol{R}_e$ or $\mathcal{Q}_e$, i.e.

$$\boldsymbol{R}_e = \boldsymbol{R}_e(\boldsymbol{q}) \tag{2.2a}$$

$$\mathcal{Q}_e = \mathcal{Q}_e(\boldsymbol{q}) \,. \tag{2.2b}$$

Let $\dot{\boldsymbol{q}}$ denote the vector of joint velocities, $\dot{\boldsymbol{p}}_e$ the (3×1) vector of end-effector *linear velocity*, and $\boldsymbol{\omega}_e$ the (3×1) vector of end-effector *angular velocity*. The *differential kinematics* model gives the relationship between $\dot{\boldsymbol{q}}$ and

$$\boldsymbol{v}_e = \begin{bmatrix} \dot{\boldsymbol{p}}_e \\ \boldsymbol{\omega}_e \end{bmatrix} \tag{2.3}$$

in the form

$$\boldsymbol{v}_e = \boldsymbol{J}(\boldsymbol{q})\dot{\boldsymbol{q}} \,, \tag{2.4}$$

where $\boldsymbol{J}$ is the $(6 \times n)$ end-effector geometric *Jacobian* matrix [63]. The Jacobian can be partitioned as

$$\boldsymbol{J} = \begin{bmatrix} \boldsymbol{J}_p \\ \boldsymbol{J}_o \end{bmatrix} \tag{2.5}$$

to separate the contributions of the joint velocities to the linear and the angular velocity in (2.3). The joint configurations at which the matrix $\boldsymbol{J}$ is not full-rank are termed *kinematic singularities*.

The *dynamic model* of the manipulator can be written in the Lagrangian form as [63]

$$\boldsymbol{B}(\boldsymbol{q})\ddot{\boldsymbol{q}} + \boldsymbol{C}(\boldsymbol{q},\dot{\boldsymbol{q}})\dot{\boldsymbol{q}} + \boldsymbol{F}\dot{\boldsymbol{q}} + \boldsymbol{g}(\boldsymbol{q}) = \boldsymbol{\tau} - \boldsymbol{J}^{\mathrm{T}}(\boldsymbol{q})\boldsymbol{h} \,, \tag{2.6}$$

where $\boldsymbol{B}(\boldsymbol{q})$ is the $n \times n$ symmetric and positive definite inertia matrix, $\boldsymbol{C}(\boldsymbol{q},\dot{\boldsymbol{q}})\dot{\boldsymbol{q}}$ is the $n \times 1$ vector of the Coriolis and centrifugal torques, $\boldsymbol{g}(\boldsymbol{q})$ is the $n \times 1$ vector of the gravitational torques, $\boldsymbol{F}\dot{\boldsymbol{q}}$ is the $n \times 1$ vector of the viscous friction torques, $\boldsymbol{\tau}$ is the $n \times 1$ vector of joint driving torques and $\boldsymbol{h} = [\,\boldsymbol{f}^{\mathrm{T}} \quad \boldsymbol{\mu}^{\mathrm{T}}\,]^{\mathrm{T}}$ is the 6×1 vector of force and torque exerted by the manipulator's end effector on the environment. Concerning the friction torques, only viscous type of friction has been considered because an accurate model of joint friction is difficult to be set up; however, complex statical and dynamical models can be found in the literature to take into account nonlinear phenomena such as Coulomb friction and Stribeck effects [5, 22].

2.2 Resolved Acceleration Control

As the present book is focused on the problem of interaction control and, as explained in the introduction, the control problem formulation will be stated at task level, in this chapter the problem of motion control will be expressed in terms of task space variables and faced through direct task space feedback.

The motion control problem for a robot manipulator can be formulated as finding the joint torques which ensure that the end effector attains a desired position and orientation. In particular, the goal is to make the end effector *track a desired position and orientation trajectory* .

Many strategies have been proposed in the literature to perform *tracking control* of a robot manipulator, but most of them can be reconducted to a classical technique, called *inverse dynamics control*, which is a particular case of the *feedback linearization* technique devised for controlling a wide class of nonlinear systems. The aim of the control law is to linearize and decouple the mechanical system by simply adding the nonlinear terms such as Coriolis and centrifugal torques, friction torques and gravity torques to the control input and then weighting the control input itself by the inertia matrix. With reference to the robot dynamic model in (2.6), it is easy to understand that with such a philosophy the joint driving torques have to be taken as

$$\boldsymbol{\tau} = \boldsymbol{B}(\boldsymbol{q})\boldsymbol{\phi} + \boldsymbol{C}(\boldsymbol{q}, \dot{\boldsymbol{q}})\dot{\boldsymbol{q}} + \boldsymbol{F}\dot{\boldsymbol{q}} + \boldsymbol{g}(\boldsymbol{q}) \,, \tag{2.7}$$

where it has been assumed $\boldsymbol{h} = \boldsymbol{0}$, i.e. no interaction with the environment, while $\boldsymbol{\phi}$ constitutes a new control input to be properly designed.

Folding the control law (2.7) into the system model (2.6), and taking into account that $\boldsymbol{B}(\boldsymbol{q})$ is always nonsingular [63], yields

$$\ddot{\boldsymbol{q}} = \boldsymbol{\phi} \,, \tag{2.8}$$

which constitutes a linear and decoupled system corresponding to a double integrator between the input $\boldsymbol{\phi}$ and the output $\boldsymbol{q}$. The quantity $\boldsymbol{\phi}$ represents a *resolved acceleration* in terms of joint variables.

Equation (2.8) has been obtained under the assumption of perfect compensation of the terms in (2.6). In case of imperfect compensation, a mismatching occurs which causes the presence of a disturbance term in (2.8), i.e.

$$\ddot{\boldsymbol{q}} = \boldsymbol{\phi} - \boldsymbol{\delta} \,. \tag{2.9}$$

In practice, the disturbance $\boldsymbol{\delta}$ is mainly due to unmodelled dynamics, such as imperfect compensation of friction torques since they are difficult to model accurately [9].

Since the output of our control system is the end-effector pose, while the output of the linearized system in (2.8) is the joint acceleration, it is necessary to compute the relationship between the joint accelerations and the end-effector linear and angular accelerations

$$\dot{\boldsymbol{v}}_e = \boldsymbol{J}(\boldsymbol{q})\ddot{\boldsymbol{q}} + \dot{\boldsymbol{J}}(\boldsymbol{q}, \dot{\boldsymbol{q}})\dot{\boldsymbol{q}} \, . \tag{2.10}$$

At this point, if a non-redundant manipulator, i.e., a $n = 6$ joint robot, and a singularity-free region of the workspace are assumed, it is possible to invert the geometric Jacobian getting the new control input $\boldsymbol{\phi}$

$$\boldsymbol{\phi} = \boldsymbol{J}^{-1}(\boldsymbol{q}) \left(\boldsymbol{a} - \dot{\boldsymbol{J}}(\boldsymbol{q}, \dot{\boldsymbol{q}})\dot{\boldsymbol{q}} \right) \, , \tag{2.11}$$

which, in view of (2.10), leads to

$$\dot{\boldsymbol{v}}_e = \boldsymbol{a} \, , \tag{2.12}$$

where $\boldsymbol{a}$ attains the meaning of a *resolved acceleration* in terms of task space variables. The redundant case ($n > 6$) will be tackled in the next chapter, and many techniques exist to handle the problem of kinematic singularities, which is not explicitly considered in this book. The interested reader is referred to [24] and references therein.

In presence of a disturbance as in (2.9), Equation (2.12) shall be modified into

$$\dot{\boldsymbol{v}}_e = \boldsymbol{a} - \boldsymbol{d} \, , \tag{2.13}$$

where $\boldsymbol{d} = \boldsymbol{J}(\boldsymbol{q})\boldsymbol{\delta}$.

In view of the partition of $\boldsymbol{v}_e$ in (2.3), it is appropriate to partition the vector $\boldsymbol{a}$ into its linear and angular components, i.e.

$$\boldsymbol{a} = \begin{bmatrix} \boldsymbol{a}_p \\ \boldsymbol{a}_o \end{bmatrix} \, , \tag{2.14}$$

where $\boldsymbol{a}_p$ and $\boldsymbol{a}_o$ are (3×1) vectors. Therefore, Equation (2.12) can be rewritten as

$$\ddot{\boldsymbol{p}}_e = \boldsymbol{a}_p \tag{2.15a}$$

$$\dot{\boldsymbol{\omega}}_e = \boldsymbol{a}_o \, , \tag{2.15b}$$

where $\boldsymbol{a}_p$ and $\boldsymbol{a}_o$ shall be designed so as to ensure tracking of the desired end-effector position and orientation trajectory, respectively.

First, consider the problem of position tracking. The desired position trajectory is specified in terms of the position vector $\boldsymbol{p}_d(t)$ of the origin of a desired frame Σ_d, linear velocity vector $\dot{\boldsymbol{p}}_d(t)$ and linear acceleration vector $\ddot{\boldsymbol{p}}_d(t)$.

A *position error* between the desired and the actual end-effector position can be defined as

$$\Delta \boldsymbol{p}_{de} = \boldsymbol{p}_d - \boldsymbol{p}_e , \tag{2.16}$$

where Δ is the operator of vector difference, and the double subscript denotes the corresponding frames. Then, the resolved linear acceleration can be chosen as

$$\boldsymbol{a}_p = \ddot{\boldsymbol{p}}_d + \boldsymbol{K}_{Dp} \Delta \dot{\boldsymbol{p}}_{de} + \boldsymbol{K}_{Pp} \Delta \boldsymbol{p}_{de} , \tag{2.17}$$

where $\boldsymbol{K}_{Dp}$ and $\boldsymbol{K}_{Pp}$ are suitable feedback matrix gains. Substituting (2.17) into (2.15a) gives the closed-loop dynamic behaviour of the position error

$$\Delta \ddot{\boldsymbol{p}}_{de} + \boldsymbol{K}_{Dp} \Delta \dot{\boldsymbol{p}}_{de} + \boldsymbol{K}_{Pp} \Delta \boldsymbol{p}_{de} = \boldsymbol{0} . \tag{2.18}$$

The system (2.18) is exponentially stable for any choice of positive definite $\boldsymbol{K}_{Dp}$ and $\boldsymbol{K}_{Pp}$, and thus tracking of $\boldsymbol{p}_d$ and $\dot{\boldsymbol{p}}_d$ is ensured.

The main challenge, now, is to extend this result to the orientation tracking problem. Many approaches have been proposed in literature, but most of them do not take into proper consideration the task geometry. In the following four subsections, different control laws will be described based on the different representations of orientation and then compared with respect to the task geometric consistency.

For this purpose, it is worth considering first the way how to express the desired orientation trajectory. In order to save consistency with the task geometry, the most natural way to express a desired rotation is to specify a desired rotation axis $\boldsymbol{r}_d$ and a desired angle of rotation $\vartheta_d(t)$ about this axis. Then, it is straightforward to compute the desired rotation matrix $\boldsymbol{R}_d(t)$ expressing the time-varying orientation of the desired frame Σ_d. To fully assign the orientation trajectory, an angular velocity $\boldsymbol{\omega}_d(t)$ and acceleration $\dot{\boldsymbol{\omega}}_d(t)$ are specified as vectors parallel to the rotation axis $\boldsymbol{r}_d$, whose magnitude are the time derivatives of the rotation angle $\vartheta_d(t)$. It should be pointed out that this method, for assigning a desired orientation trajectory for the robot end effector, is the only one which posses a clear geometrical meaning, unless the desired rotation happens about one of the principal axes of a Cartesian frame, in that case Euler angles still retain a geometrical interpretation.

2.2.1 Euler angles

In the operational space approach [45], the orientation error is computed in a form analytically analogous to (2.16), by simply describing desired and actual end-effector orientation through the two sets of three Euler angles, i.e.

$$\Delta \boldsymbol{\varphi}_{de} = \boldsymbol{\varphi}_d - \boldsymbol{\varphi}_e , \tag{2.19}$$

where $\boldsymbol{\varphi}_d$ and $\boldsymbol{\varphi}_e$ are the set of Euler angles that can be extracted respectively from the rotation matrices $\boldsymbol{R}_d$ and $\boldsymbol{R}_e$ describing the orientation of Σ_d and Σ_e.

Since the resolved acceleration $\boldsymbol{a}_o$ to be designed is related to the end-effector angular acceleration through (2.15b), it is worth considering the relationship between the time derivative of the Euler angles and the angular velocity [63]

$$\boldsymbol{\omega}_e = \boldsymbol{T}(\boldsymbol{\varphi}_e)\dot{\boldsymbol{\varphi}}_e \, , \tag{2.20}$$

where $\boldsymbol{T}$ is a transformation matrix that depends on the particular set of Euler angles considered; this matrix is singular for two values of $\boldsymbol{\varphi}_e$, the so-called representation singularities. Differentiating the previous equation with respect to time yields

$$\dot{\boldsymbol{\omega}}_e = \boldsymbol{T}(\boldsymbol{\varphi}_e)\ddot{\boldsymbol{\varphi}}_e + \dot{\boldsymbol{T}}(\boldsymbol{\varphi}_e, \dot{\boldsymbol{\varphi}}_e)\dot{\boldsymbol{\varphi}}_e \, . \tag{2.21}$$

In view of (2.21), the resolved angular acceleration based on the *Euler angles error* can be chosen as

$$\boldsymbol{a}_o = \boldsymbol{T}(\boldsymbol{\varphi}_e)\big(\ddot{\boldsymbol{\varphi}}_d + \boldsymbol{K}_{Do}\Delta\dot{\boldsymbol{\varphi}}_{de} + \boldsymbol{K}_{Po}\Delta\boldsymbol{\varphi}_{de}\big) + \dot{\boldsymbol{T}}(\boldsymbol{\varphi}_e, \dot{\boldsymbol{\varphi}}_e)\dot{\boldsymbol{\varphi}}_e \, , \tag{2.22}$$

where $\boldsymbol{K}_{Do}$ and $\boldsymbol{K}_{Po}$ are suitable feedback matrix gains. Substituting (2.22) into (2.15b) gives the closed-loop dynamic behaviour of the orientation error

$$\Delta\ddot{\boldsymbol{\varphi}}_{de} + \boldsymbol{K}_{Do}\Delta\dot{\boldsymbol{\varphi}}_{de} + \boldsymbol{K}_{Po}\Delta\boldsymbol{\varphi}_{de} = \boldsymbol{0} \, , \tag{2.23}$$

where (2.21) has been used assuming that the matrix $\boldsymbol{T}(\boldsymbol{\varphi}_e)$ is nonsingular. The system (2.23) is exponentially stable for any choice of positive definite $\boldsymbol{K}_{Do}$ and $\boldsymbol{K}_{Po}$.

From (2.22), it is easy to recognize that to compute the new control input $\boldsymbol{a}_o$, it is necessary to compute $\boldsymbol{\varphi}_d$, $\dot{\boldsymbol{\varphi}}_d$ and $\ddot{\boldsymbol{\varphi}}_d$, since the desired trajectory is expressed in terms of $\boldsymbol{R}_d$, $\boldsymbol{\omega}_d$ and $\dot{\boldsymbol{\omega}}_d$. If $\boldsymbol{\varphi}_d$ can be extracted from $\boldsymbol{R}_d$ via a standard algorithm, to compute its time derivatives, it is necessary to invert the relationships (2.20),(2.21), i.e.

$$\dot{\boldsymbol{\varphi}}_d = \boldsymbol{T}^{-1}(\boldsymbol{\varphi}_d)\boldsymbol{\omega}_d \tag{2.24}$$

$$\ddot{\boldsymbol{\varphi}}_d = \boldsymbol{T}^{-1}(\boldsymbol{\varphi}_d)\left(\dot{\boldsymbol{\omega}}_d - \dot{\boldsymbol{T}}(\boldsymbol{\varphi}_d, \dot{\boldsymbol{\varphi}}_d)\dot{\boldsymbol{\varphi}}_d\right) \, , \tag{2.25}$$

where the matrix $\boldsymbol{T}(\boldsymbol{\varphi}_d)$ has been assumed to be nonsingular.

From these equations it follows that if tracking of $\boldsymbol{\varphi}_d$ and $\dot{\boldsymbol{\varphi}}_d$ is ensured, then tracking of $\boldsymbol{R}_d$ and $\boldsymbol{\omega}_d$ is achieved.

2.2.2 Alternative Euler angles

Still in the operational space framework, if an Euler angles-based representation of orientation is of interest, a way to overcome the drawback of representation singularities affecting the previous control law is to resort to an *alternative Euler angles orientation error* [14], which is based on the rotation matrix describing the mutual orientation between Σ_d and Σ_e, i.e.

$$^{e}\boldsymbol{R}_d = \boldsymbol{R}_e^{\mathrm{T}}\boldsymbol{R}_d \tag{2.26}$$

as in (A.19), while, as in (A.26), it is

$$ {}^{e}\dot{\boldsymbol{R}}_d = \boldsymbol{S}(\Delta^e\boldsymbol{\omega}_{de})^e\boldsymbol{R}_d \,, \tag{2.27}$$

where

$$ \Delta^e\boldsymbol{\omega}_{de} = {}^e\boldsymbol{\omega}_d - {}^e\boldsymbol{\omega}_e \tag{2.28}$$

is the end-effector angular velocity error which has been referred to Σ_e.

Let $\boldsymbol{\varphi}_{de}$ denote the set of Euler angles that can be extracted from ${}^e\boldsymbol{R}_d$. Then, in view of (2.20), the angular velocity $\Delta^e\boldsymbol{\omega}_{de}$ in (2.27) is related to the time derivative of $\boldsymbol{\varphi}_{de}$ as

$$ \Delta^e\boldsymbol{\omega}_{de} = \boldsymbol{T}(\boldsymbol{\varphi}_{de})\dot{\boldsymbol{\varphi}}_{de} \,. \tag{2.29}$$

The time derivative of (2.29) gives the acceleration relationship in the form

$$ \dot{\boldsymbol{\omega}}_e = \dot{\boldsymbol{\omega}}_d - \boldsymbol{T}_e(\boldsymbol{\varphi}_{de})\ddot{\boldsymbol{\varphi}}_{de} - \dot{\boldsymbol{T}}_e(\boldsymbol{\varphi}_{de}, \dot{\boldsymbol{\varphi}}_{de})\dot{\boldsymbol{\varphi}}_{de} \,, \tag{2.30}$$

where

$$ \boldsymbol{T}_e(\boldsymbol{\varphi}_{de}) = \boldsymbol{R}_e\boldsymbol{T}(\boldsymbol{\varphi}_{de}) \,. \tag{2.31}$$

In view of (2.30), the resolved angular acceleration can be chosen as

$$ \boldsymbol{a}_o = \dot{\boldsymbol{\omega}}_d + \boldsymbol{T}_e(\boldsymbol{\varphi}_{de})(\boldsymbol{K}_{Do}\dot{\boldsymbol{\varphi}}_{de} + \boldsymbol{K}_{Po}\boldsymbol{\varphi}_{de}) - \dot{\boldsymbol{T}}_e(\boldsymbol{\varphi}_{de}, \dot{\boldsymbol{\varphi}}_{de})\dot{\boldsymbol{\varphi}}_{de} \,, \tag{2.32}$$

where $\boldsymbol{K}_{Do}$ and $\boldsymbol{K}_{Po}$ are suitable feedback matrix gains. Substituting (2.32) into (2.15b) gives the closed-loop dynamic behaviour of the orientation error

$$ \ddot{\boldsymbol{\varphi}}_{de} + \boldsymbol{K}_{Do}\dot{\boldsymbol{\varphi}}_{de} + \boldsymbol{K}_{Po}\boldsymbol{\varphi}_{de} = \boldsymbol{0} \,, \tag{2.33}$$

where (2.30) has been used assuming that the matrix $\boldsymbol{T}_e(\boldsymbol{\varphi}_{de})$ is nonsingular. The system (2.33) is exponentially stable for any choice of positive definite $\boldsymbol{K}_{Do}$ and $\boldsymbol{K}_{Po}$; convergence to $\boldsymbol{\varphi}_{de} = \boldsymbol{0}$ and $\dot{\boldsymbol{\varphi}}_{de} = \boldsymbol{0}$ is ensured, which in turn implies tracking of $\boldsymbol{R}_d$ and $\boldsymbol{\omega}_d$.

As concerns the generation of the desired trajectory, to compute the new control input $\boldsymbol{a}_o$ it is necessary to extract $\boldsymbol{\varphi}_{de}$ from ${}^e\boldsymbol{R}_d$, to calculate $\dot{\boldsymbol{\varphi}}_{de}$ and $\ddot{\boldsymbol{\varphi}}_{de}$ by inverting (2.29),(2.30). This implies that once again a greater computational burden is required to save task geometric consistency at task planning level and, at the same time, use an Euler angles-based representation. More details about computational issues will be discussed later.

At this point it should be pointed out the clear advantage of the alternative over the classical Euler angles error based on (2.19). Among the twelve possible Euler angles representations, if a representation $\boldsymbol{\varphi}_{de} = [\,\alpha_{de} \quad \beta_{de} \quad \gamma_{de}\,]^{\mathrm{T}}$, for which $\boldsymbol{T}(\boldsymbol{0})$ is nonsingular, is adopted, representation singularities occur only for large orientation errors, e.g. when $\beta_{de} = \pm\pi/2$ for the XYZ representation; in fact it can be shown that

$$ \boldsymbol{T}(\boldsymbol{\varphi}_{de}) = \begin{bmatrix} 1 & 0 & \sin\beta_{de} \\ 0 & \cos\alpha_{de} & -\sin\alpha_{de}\cos\beta_{de} \\ 0 & \sin\alpha_{de} & \cos\alpha_{de}\cos\beta_{de} \end{bmatrix} . \tag{2.34}$$

Notice that it is not advisable to choose the widely-adopted ZYZ representation which is singular right at $\boldsymbol{\varphi}_{de} = \mathbf{0}$ (see, e.g. [63]). In other words, the ill-conditioning of matrix $\boldsymbol{T}$ is not influenced by the desired or actual end-effector orientation but only by the orientation error; hence, as long as the error parameter $|\beta_{de}| < \pi/2$, the behaviour of system (2.15b) and (2.30) is not affected by representation singularities, and the control law is well defined.

2.2.3 Angle/axis

For both the above solutions of the resolved acceleration control problem, consistency with the task geometry of the control law does not hold in the sense that it is always necessary to reformulate the references for the control system loosing that consistency. Further, the problem of representation singularities has been only partially overcome thanks to the adoption of the proposed alternative Euler angles representation.

To fully preserve geometric consistency of the desired trajectory in terms of geometrically meaningful quantities, it is worth adopting a suitable representation of end-effector orientation into the design of the resolved acceleration $\boldsymbol{a}_o$ itself.

A geometrically meaningful definition of orientation error can be obtained using an angle/axis representation. In view of (2.26), the mutual orientation between Σ_d and Σ_e is described by ${}^e\boldsymbol{R}_d$, and thus the orientation error can be defined in terms of the general expression

$$ {}^e\boldsymbol{o}_{de} = f(\vartheta_{de})\,{}^e\boldsymbol{r}_{de}\,, \tag{2.35} $$

where ϑ_{de} and ${}^e\boldsymbol{r}_{de}$ are respectively the rotation and the unit vector corresponding to ${}^e\boldsymbol{R}_d$, and $f(\cdot)$ is a smooth scalar function with $f(0) = 0$. Common choices for $f(\vartheta)$ are summarized in Tab. 2.1.

Table 2.1. Common choices for $f(\vartheta)$

Representation	$f(\vartheta)$
Classical angle/axis	$\sin(\vartheta)$
Quaternion	$\sin(\vartheta/2)$
Rodrigues' parameters	$\tan(\vartheta/2)$
Simple rotation	ϑ

In the following, the classical angle/axis representation is analysed [50], while the quaternion will be treated in the next subsection. Hence, the *angle/axis orientation error* is

$$ {}^e\boldsymbol{o}'_{de} = \sin(\vartheta_{de})\,{}^e\boldsymbol{r}_{de}\,, \tag{2.36} $$

which is usually referred to the base frame, i.e.

$$\boldsymbol{o}'_{de} = \boldsymbol{R}_e {}^e\boldsymbol{o}'_{de} \, . \tag{2.37}$$

A computational expression of $\boldsymbol{o}'_{de}$ in terms of the rotation matrices representing orientation of Σ_d and Σ_e can be derived in the form [63]

$$\boldsymbol{o}'_{de} = \frac{1}{2}\left(\boldsymbol{S}(\boldsymbol{n}_e)\boldsymbol{n}_d + \boldsymbol{S}(\boldsymbol{s}_e)\boldsymbol{s}_d + \boldsymbol{S}(\boldsymbol{a}_e)\boldsymbol{a}_d\right) \, . \tag{2.38}$$

The time derivative of the orientation error (2.38) can be related to the angular velocities of Σ_d and Σ_e as

$$\dot{\boldsymbol{o}}'_{de} = \boldsymbol{L}^{\mathrm{T}}\boldsymbol{\omega}_d - \boldsymbol{L}\boldsymbol{\omega}_e \, , \tag{2.39}$$

where

$$\boldsymbol{L} = -\frac{1}{2}\left(\boldsymbol{S}(\boldsymbol{n}_d)\boldsymbol{S}(\boldsymbol{n}_e) + \boldsymbol{S}(\boldsymbol{s}_d)\boldsymbol{S}(\boldsymbol{s}_e) + \boldsymbol{S}(\boldsymbol{a}_d)\boldsymbol{S}(\boldsymbol{a}_e)\right) \, . \tag{2.40}$$

The second time derivative is

$$\ddot{\boldsymbol{o}}'_{de} = \boldsymbol{L}^{\mathrm{T}}\dot{\boldsymbol{\omega}}_d + \dot{\boldsymbol{L}}^{\mathrm{T}}\boldsymbol{\omega}_d - \boldsymbol{L}\dot{\boldsymbol{\omega}}_e - \dot{\boldsymbol{L}}\boldsymbol{\omega}_e \, . \tag{2.41}$$

In view of (2.15b),(2.41), the resolved angular acceleration can be chosen as [47]

$$\boldsymbol{a}_o = \boldsymbol{L}^{-1}\left(\boldsymbol{L}^{\mathrm{T}}\dot{\boldsymbol{\omega}}_d + \dot{\boldsymbol{L}}^{\mathrm{T}}\boldsymbol{\omega}_d - \dot{\boldsymbol{L}}\boldsymbol{\omega}_e + \boldsymbol{K}_{Do}\dot{\boldsymbol{o}}'_{de} + \boldsymbol{K}_{Po}\boldsymbol{o}'_{de}\right) \, , \tag{2.42}$$

where $\boldsymbol{K}_{Do}$ and $\boldsymbol{K}_{Po}$ are suitable feedback matrix gains, and $\boldsymbol{L}$ is nonsingular provided that the angle ϑ_{de} belongs to the interval $(-\pi/2, \pi/2)$; it can be shown that this restriction is equivalent to the conditions $\boldsymbol{n}_e^{\mathrm{T}}\boldsymbol{n}_d > 0$, $\boldsymbol{s}_e^{\mathrm{T}}\boldsymbol{s}_d > 0$, $\boldsymbol{a}_e^{\mathrm{T}}\boldsymbol{a}_d > 0$.

Substituting (2.42) into (2.15b) gives the closed-loop dynamic behaviour of the orientation error

$$\ddot{\boldsymbol{o}}'_{de} + \boldsymbol{K}_{Do}\dot{\boldsymbol{o}}'_{de} + \boldsymbol{K}_{Po}\boldsymbol{o}'_{de} = \boldsymbol{0} \, , \tag{2.43}$$

which is a linear and decoupled system analogous to the position error system (2.18) as well to the orientation error systems (2.23),(2.33). Exponential stability is guaranteed for any choice of positive definite $\boldsymbol{K}_{Do}$ and $\boldsymbol{K}_{Po}$; convergence to $\boldsymbol{o}'_{de} = \boldsymbol{0}$ and $\dot{\boldsymbol{o}}'_{de} = \boldsymbol{0}$ is ensured, which in turn implies tracking of $\boldsymbol{R}_d$ and $\boldsymbol{\omega}_d$.

Equation (2.42) reveals that the price to pay to obtain a linear and decoupled system is a large computational burden and the possible occurrence of a singularity. On the other hand, a simpler angle/axis scheme based on the error (2.38) can be devised where the resolved angular acceleration is chosen as [50]

$$\boldsymbol{a}_o = \dot{\boldsymbol{\omega}}_d + \boldsymbol{K}_{Do}\Delta\boldsymbol{\omega}_{de} + \boldsymbol{K}_{Po}\boldsymbol{o}'_{de} \, . \tag{2.44}$$

In this case, the closed-loop dynamic behaviour of the orientation error becomes

$$\Delta\dot{\boldsymbol{\omega}}_{de} + \boldsymbol{K}_{Do}\Delta\boldsymbol{\omega}_{de} + \boldsymbol{K}_{Po}\boldsymbol{o}'_{de} = \boldsymbol{0} \, . \tag{2.45}$$

Differently from all the previous cases (2.18), (2.23), (2.33) and (2.43), the error system is nonlinear, and thus a Lyapunov argument is invoked below to ascertain its stability. To this purpose, in view of (A.16a),(A.16b), the orientation error (2.36) can be expressed in terms of a quaternion as

$$ {}^{e}\boldsymbol{o}'_{de} = 2\eta_{de}\,{}^{e}\boldsymbol{\epsilon}_{de} \,, \tag{2.46}$$

where $\{\eta_{de}, {}^{e}\boldsymbol{\epsilon}_{de}\}$ can be extracted from (2.26). Furthermore, the feedback gains are taken as scalar matrices, i.e. $\boldsymbol{K}_{Do} = k_{Do}\boldsymbol{I}$ and $\boldsymbol{K}_{Po} = k_{Po}\boldsymbol{I}$, where $\boldsymbol{I}$ denotes the (3×3) identity matrix. Let

$$ \mathcal{V} = 2k_{Po}\,{}^{e}\boldsymbol{\epsilon}_{de}^{\mathrm{T}}\,{}^{e}\boldsymbol{\epsilon}_{de} + \frac{1}{2}\Delta\boldsymbol{\omega}_{de}^{\mathrm{T}}\Delta\boldsymbol{\omega}_{de} \tag{2.47}$$

be a positive definite Lyapunov function candidate. The time derivative of (2.47) along the trajectories of system (2.45) is

$$\begin{aligned} \dot{\mathcal{V}} &= 4k_{Po}\,{}^{e}\boldsymbol{\epsilon}_{de}^{\mathrm{T}}\,{}^{e}\dot{\boldsymbol{\epsilon}}_{de} + \Delta\boldsymbol{\omega}_{de}^{\mathrm{T}}\Delta\dot{\boldsymbol{\omega}}_{de} \\ &= 2k_{Po}\,{}^{e}\boldsymbol{\epsilon}_{de}^{\mathrm{T}}\boldsymbol{E}(\eta_{de}, {}^{e}\boldsymbol{\epsilon}_{de})\Delta^{e}\boldsymbol{\omega}_{de} - k_{Do}\Delta\boldsymbol{\omega}_{de}^{\mathrm{T}}\Delta\boldsymbol{\omega}_{de} - 2k_{Po}\eta_{de}\Delta^{e}\boldsymbol{\omega}_{de}^{\mathrm{T}}\,{}^{e}\boldsymbol{\epsilon}_{de} \\ &= -k_{Do}\Delta\boldsymbol{\omega}_{de}^{\mathrm{T}}\Delta\boldsymbol{\omega}_{de} \,, \end{aligned} \tag{2.48}$$

where the propagation rule based on (A.28a),(A.28b) has been exploited, i.e.

$$ \dot{\eta}_{de} = -\frac{1}{2}\,{}^{e}\boldsymbol{\epsilon}_{de}^{\mathrm{T}}\Delta^{e}\boldsymbol{\omega}_{de} \tag{2.49a}$$

$$ {}^{e}\dot{\boldsymbol{\epsilon}}_{de} = \frac{1}{2}\boldsymbol{E}(\eta_{de}, {}^{e}\boldsymbol{\epsilon}_{de})\Delta^{e}\boldsymbol{\omega}_{de} \tag{2.49b}$$

with $\boldsymbol{E}$ defined as in (A.29).

Since $\dot{\mathcal{V}}$ is only negative semi-definite, in view of LaSalle theorem [44], the system asymptotically converges to the invariant set described by the following equilibria:

$$ \mathcal{E}_1 = \{\eta_{de} = 1,\ \ {}^{e}\boldsymbol{\epsilon}_{de} = \boldsymbol{0},\ \ \Delta\boldsymbol{\omega}_{de} = \boldsymbol{0}\} \tag{2.50a}$$

$$ \mathcal{E}_2 = \{\eta_{de} = -1,\ \ {}^{e}\boldsymbol{\epsilon}_{de} = \boldsymbol{0},\ \ \Delta\boldsymbol{\omega}_{de} = \boldsymbol{0}\} \tag{2.50b}$$

$$ \mathcal{E}_3 = \{\eta_{de} = 0,\ \ {}^{e}\boldsymbol{\epsilon}_{de} : \|{}^{e}\boldsymbol{\epsilon}_{de}\| = 1,\ \ \Delta\boldsymbol{\omega}_{de} = \boldsymbol{0}\}. \tag{2.50c}$$

The equilibria in the set $\mathcal{E}_3$ are all unstable. To see this, consider (2.47) which, in view of (2.48), is a decreasing function. At any equilibrium in (2.50c), it is

$$ \mathcal{V}_{\infty} = 2k_{Po} \,. \tag{2.51}$$

Take a small perturbation $\eta_{de} = \sigma$ around such equilibrium; then, from (A.15) it follows ${}^{e}\boldsymbol{\epsilon}_{de}^{\mathrm{T}}\,{}^{e}\boldsymbol{\epsilon}_{de} = 1 - \sigma^2$. The perturbed Lyapunov function is

$$ \mathcal{V}_{\sigma} = 2k_{Po} - 2\sigma^2 k_{Po} < \mathcal{V}_{\infty} \tag{2.52}$$

and thus, since (2.47) is decreasing, $\mathcal{V}$ will never return to $\mathcal{V}_{\infty}$, implying that the equilibria in $\mathcal{E}_3$ are unstable. Therefore, the system asymptotically converges to either $\mathcal{E}_1$ or $\mathcal{E}_2$; since both quaternions for those equilibria represent the same orientation (see Appendix A for more details), it can be concluded that tracking of $\boldsymbol{R}_d$ and $\boldsymbol{\omega}_d$ is achieved.

It is worth emphasizing that, compared to the previous angle/axis scheme based on (2.42), the restriction on ϑ_{de} does no longer hold. Nevertheless, for large values of ϑ_{de}, the angle/axis scheme based on (2.44) may lead to alignment of Σ_d and Σ_e with $\vartheta_{de} = \pm 2\pi$ as in the equilibrium $\mathcal{E}_2$; such occurrence is of no practical interest since ϑ_{de} is typically small for the tracking control at issue. In any case, it should be pointed out that the computation of the orientation error $\boldsymbol{o}_{de}$ and consequently of the resolved acceleration $\boldsymbol{a}_o$ requires the computation of the rotation matrix $\boldsymbol{R}_d$ starting from the assigned orientation axis $\boldsymbol{r}_d$ and the assigned rotation angle $\vartheta_d(t)$, with the the subsequent computational load. On the other hand, it would be useful to compute the orientation error only through the references, avoiding the occurrence of any singularity. This will be the topic of the next section.

2.2.4 Quaternion

Since the geometrically consistent desired trajectory is assigned in terms of an angle/axis representation and for all the representations seen in the previous section the computation of $\boldsymbol{a}_o$ passed through a rotation matrix, it is necessary to resort to a description of orientation, which could completely substitute the rotation matrices and, at the same time, would have a geometrical meaning.

The only way to achieve this characteristic is to analyse the properties of the set of rotations, i.e., the group $SO(3)$. In Appendix A, a detailed discussion about this issue is carried out, and the main conclusion is that only non-minimal representation of orientation can fully substitute rotation matrices. In this class of descriptions of rotations, the *unit quaternion* is a very compact one (only four parameters) and it has a clear geometrical meaning, since it can be described as a special angle/axis representation. However, on the group of unit quaternions it is possible to define a complete algebra [26] and, owing to that, every operation with rotation matrices is possible as with the corresponding unit quaternions. This kind of representation has been successfully used for the attitude control problem of rigid bodies (spacecraft) in [69, 33].

If $\mathcal{Q}_e$ and $\mathcal{Q}_d$ represent the unit quaternions describing the orientation of Σ_e and Σ_d with respect to the base frame, the mutual orientation of the two frames is

$$\mathcal{Q}_{de} = \mathcal{Q}_e^{-1} * \mathcal{Q}_d \,. \tag{2.53}$$

By taking into account (A.16a),(A.16b), it can be recognized that the vector part of this quaternion can be written as

$${}^e\boldsymbol{\epsilon}_{de} = \sin\frac{\vartheta_{de}}{2}\,{}^e\boldsymbol{r}_{de}\,, \tag{2.54}$$

and thus, it possible to define an orientation error belonging to the class of angle/axis type, as

$$\boldsymbol{o}''_{de} = {}^e\boldsymbol{\epsilon}_{de} . \tag{2.55}$$

Notice that the computation of the rotation matrix $\boldsymbol{R}_d$ from $\boldsymbol{r}_d$ and ϑ_d is not needed to compute the orientation error $\boldsymbol{o}''_{de}$ because, in view of the quaternion interpretation as *Euler parameters* (see Appendix A), computation of the unit quaternion $\mathcal{Q}_d$ from $\boldsymbol{r}_d$ and θ_d is straightforward.

The relationship between the time derivative of the orientation error $\boldsymbol{o}''_{de}$ and the angular velocity error in (2.28) is established by (2.49b).

The resolved angular acceleration based on the *quaternion orientation error* can be chosen as [73]

$$\boldsymbol{a}_o = \dot{\boldsymbol{\omega}}_d + \boldsymbol{K}_{Do}\Delta\boldsymbol{\omega}_{de} + \boldsymbol{K}_{Po}\boldsymbol{R}_e{}^e\boldsymbol{\epsilon}_{de} , \tag{2.56}$$

where $\boldsymbol{K}_{Do}$ and $\boldsymbol{K}_{Po}$ are suitable feedback matrix gains, and the orientation error has been referred to the base frame. Substituting (2.56) into (2.15b) gives the closed-loop dynamic behaviour of the orientation error

$$\Delta\dot{\boldsymbol{\omega}}_{de} + \boldsymbol{K}_{Do}\Delta\boldsymbol{\omega}_{de} + \boldsymbol{K}_{Po}\boldsymbol{R}_e{}^e\boldsymbol{\epsilon}_{de} = \boldsymbol{0} . \tag{2.57}$$

Similarly to the angle/axis case above, a Lyapunov argument is invoked below to ascertain stability of system (2.57). Again, the feedback gains are taken as scalar matrices, i.e. $\boldsymbol{K}_{Do} = k_{Do}\boldsymbol{I}$ and $\boldsymbol{K}_{Po} = k_{Po}\boldsymbol{I}$. Let

$$\mathcal{V} = k_{Po}\left((\eta_{de} - 1)^2 + {}^e\boldsymbol{\epsilon}_{de}^{\mathrm{T}}\,{}^e\boldsymbol{\epsilon}_{de}\right) + \frac{1}{2}\Delta\boldsymbol{\omega}_{de}^{\mathrm{T}}\Delta\boldsymbol{\omega}_{de} \tag{2.58}$$

be a positive definite Lyapunov function candidate. The time derivative of (2.58) along the trajectories of system (2.57) is

$$\begin{aligned}
\dot{\mathcal{V}} &= 2k_{Po}\left((\eta_{de} - 1)\dot{\eta}_{de} + {}^e\boldsymbol{\epsilon}_{de}^{\mathrm{T}}\,{}^e\dot{\boldsymbol{\epsilon}}_{de}\right) + \Delta\boldsymbol{\omega}_{de}^{\mathrm{T}}\Delta\dot{\boldsymbol{\omega}}_{de} \\
&= k_{Po}\left(-(\eta_{de} - 1){}^e\boldsymbol{\epsilon}_{de}^{\mathrm{T}}\Delta{}^e\boldsymbol{\omega}_{de} + {}^e\boldsymbol{\epsilon}_{de}^{\mathrm{T}}\boldsymbol{E}(\eta_{de}, {}^e\boldsymbol{\epsilon}_{de})\Delta{}^e\boldsymbol{\omega}_{de}\right) \\
&\quad - k_{Do}\Delta\boldsymbol{\omega}_{de}^{\mathrm{T}}\Delta\boldsymbol{\omega}_{de} - k_{Po}\Delta{}^e\boldsymbol{\omega}_{de}^{\mathrm{T}}\,{}^e\boldsymbol{\epsilon}_{de} \\
&= -k_{Do}\Delta\boldsymbol{\omega}_{de}^{\mathrm{T}}\Delta\boldsymbol{\omega}_{de}
\end{aligned} \tag{2.59}$$

where (2.49a),(2.49b) have been exploited.

Since $\dot{\mathcal{V}}$ is only negative semi-definite, in view of LaSalle theorem, the system asymptotically converges to the invariant set described by the two equilibria $\mathcal{E}_1$ and $\mathcal{E}_2$ in (2.50a),(2.50b).

The equilibrium $\mathcal{E}_2$ is unstable. To see this, consider (2.58) which, in view of (2.59), is a decreasing function. At the equilibrium in (2.50b), it is

$$\mathcal{V}_\infty = 4k_{Po} . \tag{2.60}$$

Take a small perturbation $\eta_{de} = -1 + \sigma$ around the equilibrium with $\sigma > 0$; then, it is ${}^e\boldsymbol{\epsilon}_{de}^{\mathrm{T}}\,{}^e\boldsymbol{\epsilon}_{de} = 2\sigma - \sigma^2$. The perturbed Lyapunov function is

$$\mathcal{V}_\sigma = 4k_{Po} - 2\sigma k_{Po} < \mathcal{V}_\infty \tag{2.61}$$

and thus, since (2.58) is decreasing, $\mathcal{V}$ will never return to $\mathcal{V}_\infty$, implying that $\mathcal{E}_2$ is unstable. Therefore, the system must asymptotically converge to $\mathcal{E}_1$, which in turn implies that tracking of $\boldsymbol{R}_d$ and $\boldsymbol{\omega}_d$ is achieved.

2.3 Comparison

In this section the main differences between the different resolved acceleration control laws will be summarized to the purpose of getting more insight into the problem of choice of the most suitable representation of orientation. The comparison will be performed from the computational point of view, from the point of view of range of applicability of the control law and, in the experimental section, from the performance point of view.

2.3.1 Computational issues

Assume that the robot manipulator is controlled in a real-time fashion, and thus the planning of the desired end-effector trajectory shall be updated on the basis of sensory information about the surrounding environment where the robot operates. Therefore, the two key elements in the analysis are the trajectory generation and the computation of the resolved acceleration $\boldsymbol{a}_o$, while the rest of the control algorithm is the same for each scheme.

As already discussed in a previous section, the way to specify a desired orientation trajectory consistent with the geometry of the task is the angle/axis method. Therefore, it is assumed that the trajectory is generated in terms of $\boldsymbol{r}_d$ and $\vartheta_d(t)$, where the time varying angle is generated according to an interpolating polynomial of 5-th order.

Regarding the actual orientation of Σ_e, this is typically available from the direct kinematics equation in terms of the rotation matrix $\boldsymbol{R}_e$ which can be computed from the joint position measurements via the direct kinematic equation (2.2a); further, the actual end-effector angular velocity $\boldsymbol{\omega}_e$ can be computed from the joint velocity measurements via (2.4).

The computational burden of the schemes presented in the previous subsections has been evaluated in terms of the number of floating-point operations and transcendental functions needed to compute the resolved angular acceleration $\boldsymbol{a}_o$. The angle/axis scheme based on (2.42) has been ruled out in view of its inherent computational complexity, and thus the scheme based on (2.44) has been considered hereafter. The results are reported in Tab. 2.2, where the additional computations needed for desired trajectory generation by those schemes not using $\boldsymbol{r}_d$, ϑ_d, $\boldsymbol{\omega}_d$ and $\dot{\boldsymbol{\omega}}_d$ are evidenced; the computations have been optimized whenever possible, e.g. by avoiding multiplications by zero and carrying out partial factorizations.

It can be recognized that the extraction of Euler angles augments the overall load for both schemes in the operational space, compared to the quaternion and the angle/axis scheme. Moreover, for all the schemes except for the

Table 2.2. Computational load of the angular part for the resolved acceleration control schemes

	Resolved acceleration		*Trajectory generation*	
Orientation error	*Flops*	*Funcs*	*Flops*	*Funcs*
Classical Euler angles	68	8	52	8
Alternative Euler angles	136	8	0	0
Angle/axis	55	0	0	0
Quaternion	60	1	21	1

quaternion-based one, the required computation of the rotation matrix $\boldsymbol{R}_d$ from the desired angle and axis increases the computational load. However, for the quaternion-based scheme it is necessary to extract the quaternion corresponding to the rotation matrix describing the actual end-effector orientation, which explains the difference between this algorithm and that based on the angle/axis representation.

As a confirmation a test for the estimation of the total computational time has been carried out on the available hardware; it amounts to: 0.235 ms for the controller based on (2.22), 0.205 ms for the controller based on (2.32), 0.155 ms for the controller based on (2.44), and 0.170 ms for the controller based on (2.56). Notice that, with reference to the data in Tab. 2.2, these times are also inclusive of the inverse dynamics computation, the basic trajectory generation and the manipulator kinematics computation, and of course the computation of the resolved linear acceleration.

2.3.2 Experiments

The feedback gains of the above controllers have been set to

$$\boldsymbol{K}_{Vp} = 75\boldsymbol{I}, \quad \boldsymbol{K}_{Pp} = 2500\boldsymbol{I}$$

for the translational part of the control law based on (2.17), to

$$\boldsymbol{K}_{Vo} = 75\boldsymbol{I}, \quad \boldsymbol{K}_{Po} = 2500\boldsymbol{I}$$

for the rotational part of the control law based on (2.22),(2.32),(2.44), while

$$\boldsymbol{K}_{Vo} = 75\boldsymbol{I}, \quad \boldsymbol{K}_{Po} = 5000\boldsymbol{I}$$

for the rotational part of the control law based on quaternions (2.56). In order to carry out a fair comparison, the above values have been chosen so as to assign the same dynamic behaviour to the various closed-loop error systems, where the nonlinear equations have been linearized for small orientation errors.

A case study has been developed to analyse the tracking performance of the various schemes. The end-effector desired position is required to make a straight line displacement of $(0.5, -0.6, 0.5)$ m along the coordinate axes of the base frame. The trajectory along the path is generated according to a 5-th order interpolating polynomial with null initial and final velocities and accelerations, and a duration of 5 s. The end-effector desired orientation is required to make a rotation of 3 rad about the axis $(0.5639, 0.5840, -0.5840)$ with respect to the base frame. The trajectory is generated according to the equivalent angle/axis method, where the axis is fixed and the angle is interpolated by a 5-th order polynomial with null initial and final velocities and accelerations, and a duration of 5 s. The initial end-effector pose has been matched with the desired one.

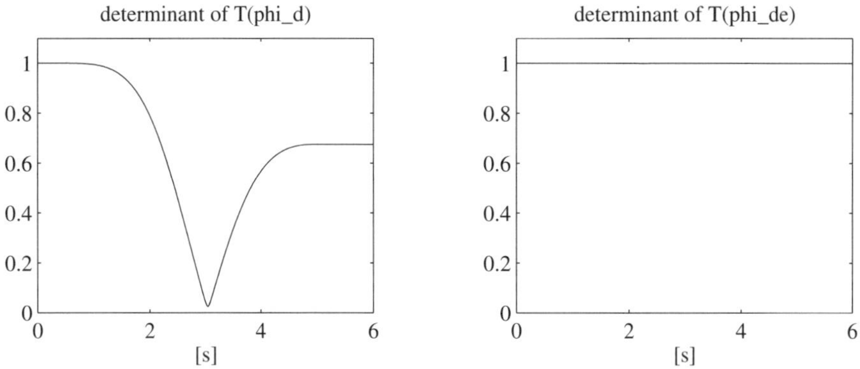

Fig. 2.1. Time histories of determinant of $\boldsymbol{T}(\boldsymbol{\varphi}_d)$ and determinant of $\boldsymbol{T}(\boldsymbol{\varphi}_{de})$

As shown on the left-hand side of Fig. 2.1, the desired end-effector orientation trajectory passes in the neighbourhood of a representation singularity for the matrix $\boldsymbol{T}(\boldsymbol{\varphi}_d)$; this is a very demanding task in the face of the typical capability of a conventional industrial robot control unit.

The results are reported in Figs. 2.2 and 2.3 for the two Euler angles feedback schemes, the classical one and the alternative one proposed in this paper. The figures illustrate the time histories of the norm of the end-effector position error $\Delta\boldsymbol{p}_{de}$, the norm of an end-effector orientation error computed as the largest singular value of the matrix $(\boldsymbol{I} - \boldsymbol{R}_d^e)$, the norm of the linear velocity error $\Delta\dot{\boldsymbol{p}}_{de}$, the norm of the angular velocity error $\Delta\boldsymbol{\omega}_{de}$, and the joint reference motor currents (expressed in DAC units). Note that the choice of the above orientation error has been motivated by the desire to refer to a single measure of tracking performance for the various schemes, no matter what type of orientation feedback has been used.

It can be recognized that the performance in terms of the position and orientation errors is good for both schemes. The steady-state errors are non-

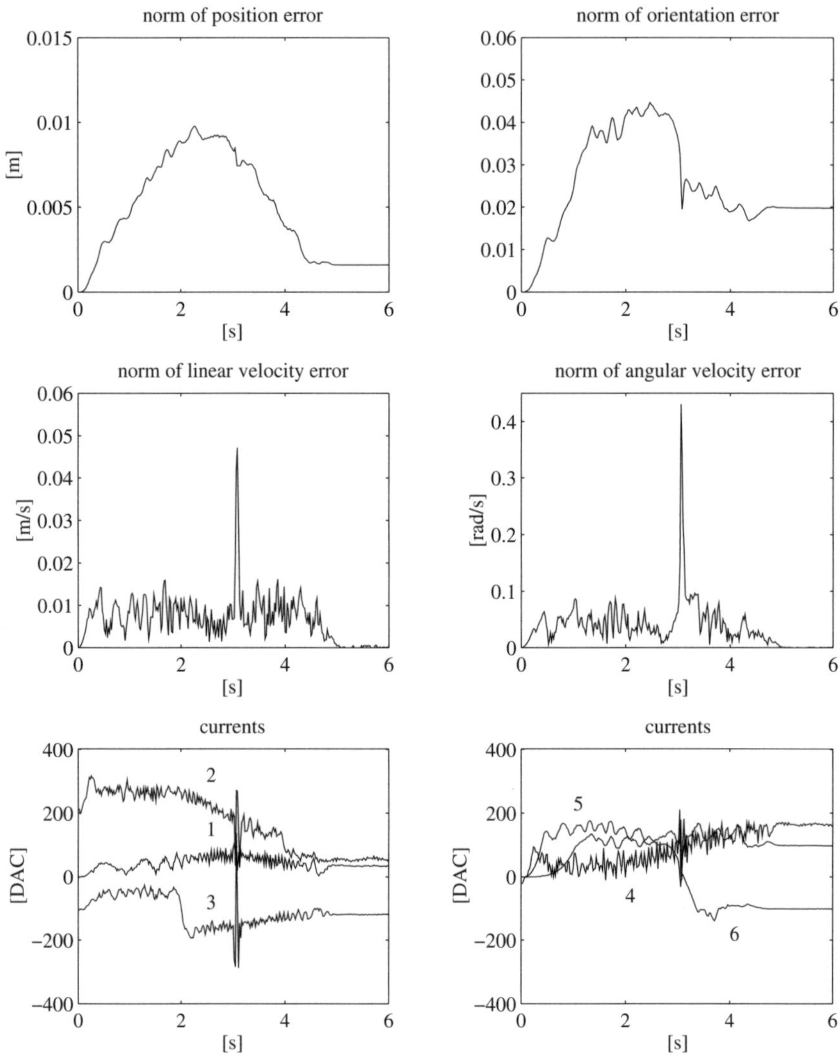

Fig. 2.2. Tracking performance of resolved acceleration control based on classical Euler angles

null because of the unavoidable imperfect compensation of static friction. A degradation of performance is observed in the linear and angular velocity errors for the scheme based on the classical Euler angles feedback; large peaks occur which are reflected also at the level of the motor currents. This phenomenon can be clearly ascribed to the closeness to the representation singularity for the end-effector orientation in correspondence of the time instant when the determinant of $\boldsymbol{T}(\boldsymbol{\varphi}_d)$ approaches zero. It should be remarked that its effect is visible not only on the angular velocity error but also on the linear velocity error, through the typical kinematic coupling between position and orientation in the non-spherical wrist manipulator. On the other hand, the scheme based on the alternative Euler angles feedback does not suffer from the occurrence of representation singularities, as confirmed on the right-hand side of Fig. 2.1 illustrating the determinant of the matrix $\boldsymbol{T}(\boldsymbol{\varphi}_{de})$ which is nearly equal to one in view of the small orientation tracking error along the trajectory.

The overall tracking performance obtained with the quaternion feedback and the angle/axis feedback schemes is practically the same as that with the alternative Euler angles feedback scheme, and thus the numerical results have not been reported for brevity.

In conclusion, a critical discussion is in order concerning the pros and cons of each scheme.

At first sight, the classical Euler angles scheme might seem the simplest one in view of its similarity with the position scheme. Nevertheless, the analysis has revealed that, besides the heavy computational load due to Euler angles extraction, there is no guarantee to avoid the occurrence of representation singularities even when good end-effector orientation tracking is achieved. On the other hand, the effort to plan a singularity-free orientation trajectory is considerable especially when the angle/axis method is adopted for meaningful task specification purposes; furthermore that is even impossible when the manipulator interacts with the environment, since the trajectory is not known a priori, because it depends on the contact force and moment.

The alternative Euler angles scheme has the main merit to almost overcome the above drawback of representation singularities, since it keenly operates on the set of Euler angles which is extracted from a single rotation matrix describing the mutual orientation between Σ_d and Σ_e. It may suffer only in the case of large orientation errors, but there is no practical worry for a convergent algorithm with matched initial conditions between the desired and the actual end-effector orientation. A weakness, however, is that the computational burden is still considerable, especially due to the required computation of the matrix $\boldsymbol{T}_e$ and its time derivatives.

The breakthrough of the quaternion scheme stands in its applicability for any magnitude of the orientation error, since it is inherently based on a singularity-free representation of orientation. Its tracking performance is apparently as good as the alternative Euler angles scheme, although the closed-loop orientation error dynamics is nonlinear. A further advantage is represented by the contained computational burden, even though the orientation error is not directly based on the desired and actual rotation matrices.

Finally, the angle/axis scheme has the least computational load among all the schemes. Its performance is worse than that of the quaternion scheme in the case of large orientation errors, whereas both schemes exhibit the same good behaviour for small orientation errors.

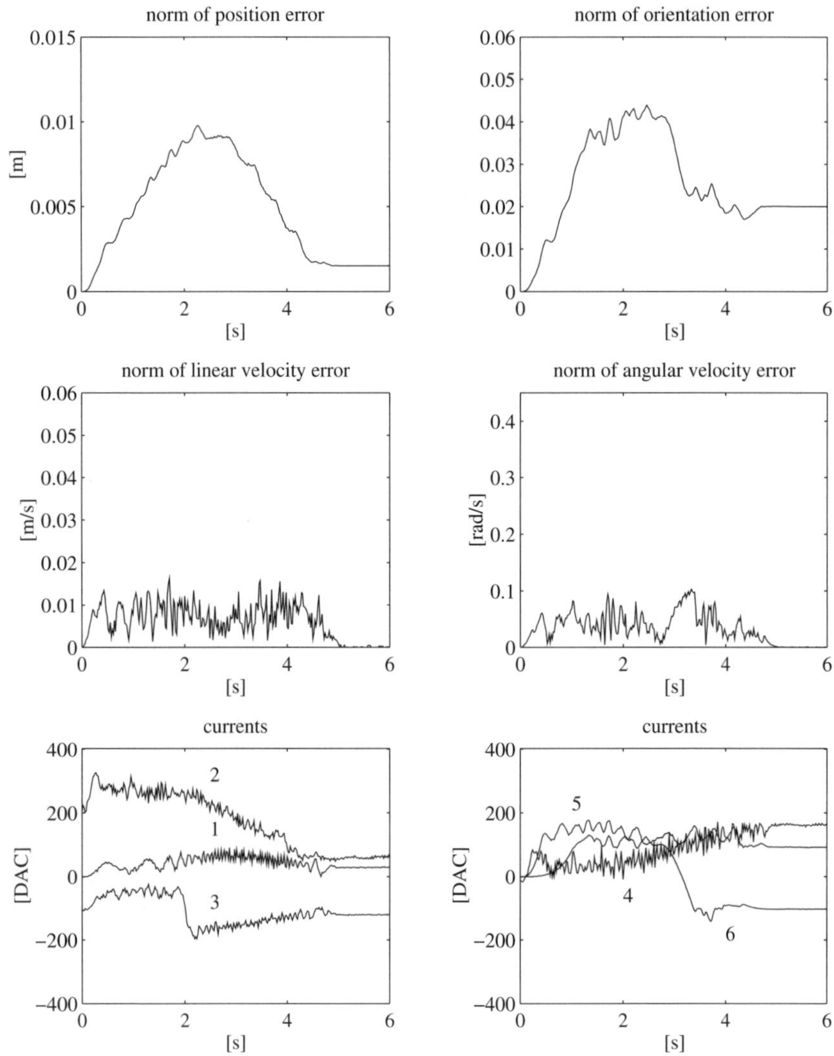

Fig. 2.3. Tracking performance of resolved acceleration control based on alternative Euler angles

3. Task Space Impedance Control

In this chapter, a first control strategy to manage the interaction of a single robot manipulator with the environment is examined, namely impedance control. First, impedance equations in the operational space are presented and then compared with the task space approach, both for translational and rotational motions. The more general case of a redundant robot is also considered and experimental results are presented to illustrate the beneficial effects of the impedance control during interaction with the environment.

3.1 Indirect Force Control

Impedance control belongs to the category of indirect force control because it does not allow the user to specify a desired contact force. In detail, the strategy conferring a compliant behaviour to the robotic system, in order to manage the interaction of the manipulator with the environment is conceived as follows.

By using an external force/torque sensor the contact force measurements are suitably used in the control system to impose a dynamic behaviour to the robotic arm which can be described by a *mechanical impedance*; namely, a dynamic relationship between contact force and end-effector displacements, characterized by a generalized *mass matrix*, a *damping matrix* and a *stiffness matrix*. It is not difficult to understand that, when the robot interacts with the environment, which is assumed to be a passive compliant environment, the dynamics of the overall system is influenced also by the characteristics of the contact surface and by the contact geometry. The idea is to design the parameters of the assigned impedance so as to have a compliant and well-damped dynamic behaviour during the entire execution of the interaction task, i.e., during the approach phase and during the contact phase.

3.2 Impedance Equation in the Operational Space

Impedance control was first proposed in the seminal work [39] and further developed by many researchers, but most of them did not place too much stress on the problem of the rotational part of the mechanical impedance; in fact, most of the interaction tasks considered in the literature were only three-DOF tasks, as defined in the introduction of this book.

To formally define a mechanical impedance, besides the *actual end-effector frame* Σ_e and the *desired end-effector frame* Σ_d, it is worth introducing the *compliant frame* Σ_c. The last frame describes position and orientation of the end effector when it is in contact with the environment, so that the relationship between the end-effector displacements and the contact force and moment is governed by the above mentioned specified impedance parameters. In other words, during the task execution, when the end effector moves in free space, it shall follow the desired trajectory whereas, when the end effector is in contact with the environment, it shall follow the compliant trajectory.

The first work dealing with a general six-DOF impedance controller was [45], but as in the simple motion control problem, the rotational part of the impedance controller was obtained by a simple analytical extension of the translational part, thanks to the use of Euler angles to represent the end-effector orientation.

In order to comprehend the main characteristics of the operational space approach and to identify its drawbacks and limits, a brief overview of the formulation will be carried out in the following subsections.

3.2.1 Translational impedance

To define the impedance equation, an energy-based argument can be pursued to give a physical interpretation of the equation. First, let

$$\Delta \boldsymbol{p}_{dc} = \boldsymbol{p}_d - \boldsymbol{p}_c \tag{3.1}$$

be the translational displacement between the origins of the two frames Σ_d and Σ_c; it is possible to define the following *translational pseudo-kinetic energy*

$$\mathcal{T}_p = \frac{1}{2}\Delta\dot{\boldsymbol{p}}_{dc}^{\mathrm{T}}\boldsymbol{M}_p\,,\Delta\dot{\boldsymbol{p}}_{dc} \tag{3.2}$$

where $\boldsymbol{M}_p$ is the symmetric and positive definite *translational mass matrix* and $\Delta\dot{\boldsymbol{p}}_{dc}$ is the time derivative of $\Delta\boldsymbol{p}_{dc}$ in (3.1), i.e., the relative velocity of the frame Σ_d with respect to the frame Σ_c, expressed in the base frame. Notice that $\mathcal{T}_p$ is a true kinetic energy only if $\boldsymbol{p}_d$ is constant and $\boldsymbol{M}_p$ is a scalar matrix.

Then consider the *potential energy* of a three-DOF spring with *translational stiffness matrix* $\boldsymbol{K}_p$ [49] (symmetric and positive definite) and equilibrium position $\boldsymbol{p}_d$, i.e.

$$\mathcal{U}_p = \frac{1}{2}\Delta\boldsymbol{p}_{dc}^{\mathrm{T}}\boldsymbol{K}_p\Delta\boldsymbol{p}_{dc}\,. \tag{3.3}$$

Now, it is possible to derive the various terms of the impedance equation by considering the associated powers of the corresponding energy terms. In detail, taking the time derivative of (3.2) yields

$$\dot{\mathcal{T}}_p = \boldsymbol{f}_I^{\mathrm{T}} \Delta\dot{\boldsymbol{p}}_{dc} , \tag{3.4}$$

where

$$\boldsymbol{f}_I = \boldsymbol{M}_p \Delta\ddot{\boldsymbol{p}}_{dc} \tag{3.5}$$

is the inertial force.

With reference to (3.3), taking the time derivative of $\mathcal{U}_p$ yields

$$\dot{\mathcal{U}}_p = \boldsymbol{f}_E^{\mathrm{T}} \Delta\dot{\boldsymbol{p}}_{dc} , \tag{3.6}$$

where

$$\boldsymbol{f}_E = \boldsymbol{K}_p \Delta\boldsymbol{p}_{dc} \tag{3.7}$$

is the elastic force; in fact the product of the elastic force and the velocity $\Delta\dot{\boldsymbol{p}}_{dc}$ is the elastic power.

Finally, a dissipative contribution can be added as

$$\boldsymbol{f}_D = \boldsymbol{D}_p \Delta\dot{\boldsymbol{p}}_{dc} , \tag{3.8}$$

where $\boldsymbol{D}_p$ is the symmetric and positive definite *translational damping matrix*.

Thus, by adding the contributions (3.5),(3.7) and (3.8), the *translational impedance equation* can be defined as

$$\boldsymbol{M}_p \Delta\ddot{\boldsymbol{p}}_{dc} + \boldsymbol{D}_p \Delta\dot{\boldsymbol{p}}_{dc} + \boldsymbol{K}_p \Delta\boldsymbol{p}_{dc} = \boldsymbol{f} , \tag{3.9}$$

where $\boldsymbol{f}$ is the force exerted by the end effector on the environment.

In order to ensure the end-effector compliant behaviour to be correct for the execution of the interaction task, the selection of the stiffness matrix plays a key role. Therefore, it is worth analysing the elastic term from a geometric point of view.

The stiffness matrix $\boldsymbol{K}_p$ can be decomposed as

$$\boldsymbol{K}_p = \boldsymbol{U}_p \boldsymbol{\Gamma}_p \boldsymbol{U}_p^{\mathrm{T}} , \tag{3.10}$$

where $\boldsymbol{\Gamma}_p = \mathrm{diag}\{\gamma_{p1}, \gamma_{p2}, \gamma_{p3}\}$ and $\boldsymbol{U}_p = [\, \boldsymbol{u}_{p1} \quad \boldsymbol{u}_{p2} \quad \boldsymbol{u}_{p3} \,]$ are respectively the eigenvalue matrix and the (orthogonal) eigenvector matrix. Then, considering a position displacement of length λ along the i-th eigenvector leads to an elastic force

$$\boldsymbol{f}_E = \boldsymbol{K}_p \Delta\boldsymbol{p}_{dc} = \gamma_{pi} \lambda \boldsymbol{u}_{pi} , \tag{3.11}$$

which represents an elastic force along the same $\boldsymbol{u}_{pi}$ axis. This implies that the translational stiffness matrix can be expressed in terms of three parameters γ_{pi} representing the stiffness along *three principal axes* $\boldsymbol{u}_{pi}$ [35], and in turn it establishes the property of *task geometric consistency* for the elastic force in (3.11).

3.2.2 Rotational impedance based on Euler angles

The above physical and geometrical interpretation of the different terms in the translational impedance equation greatly simplifies the impedance parameters selection depending on the task geometry; this is the main motivation of the geometric approach, which will be followed to define the rotational impedance equation in the task space impedance control, presented in this book and originally proposed in [16]. On the contrary, in the operational space approach the rotational impedance equation is derived via a formal analogy with the translational impedance equation, by resorting to the representation of end-effector orientation based on Euler angles.

Therefore, with reference to (3.9), the *rotational impedance equation* can be written in the form [45]

$$\boldsymbol{M}_o \Delta\ddot{\boldsymbol{\varphi}}_{dc} + \boldsymbol{D}_o \Delta\dot{\boldsymbol{\varphi}}_{dc} + \boldsymbol{K}_o \Delta\boldsymbol{\varphi}_{dc} = \boldsymbol{T}^{\mathrm{T}}(\boldsymbol{\varphi}_c)\boldsymbol{\mu} \,, \tag{3.12}$$

where $\boldsymbol{M}_o$, $\boldsymbol{D}_o$ and $\boldsymbol{K}_o$ are symmetric and positive definite matrices describing the generalized inertia, rotational damping, rotational stiffness, respectively, and

$$\Delta\boldsymbol{\varphi}_{dc} = \boldsymbol{\varphi}_d - \boldsymbol{\varphi}_c \tag{3.13}$$

is the *Euler angles orientation displacement.*

Notice the presence of the transformation matrix $\boldsymbol{T}$, necessary to transform the physical moment $\boldsymbol{\mu}$ into a quantity consistent with the terms in the first member of (3.12). More precisely, to fix the ideas, consider the elastic contribution. Let $\boldsymbol{\mu}_E$ be the elastic contribution to the physical moment, then the physical elastic power is

$$\dot{\mathcal{U}}_o = \boldsymbol{\mu}_E^{\mathrm{T}} \Delta\boldsymbol{\omega}_{dc} \,, \tag{3.14}$$

while the elastic power associated to the elastic term $\boldsymbol{K}_o \Delta\boldsymbol{\varphi}_{dc}$ is the scalar product $\Delta\boldsymbol{\varphi}_{dc}^{\mathrm{T}} \boldsymbol{K}_o \Delta\dot{\boldsymbol{\varphi}}_{dc}$; hence, to save the physical meaning, it shall be

$$\boldsymbol{\mu}_E^{\mathrm{T}} \Delta\boldsymbol{\omega}_{dc} = \Delta\boldsymbol{\varphi}_{dc}^{\mathrm{T}} \boldsymbol{K}_o \Delta\dot{\boldsymbol{\varphi}}_{dc} \,. \tag{3.15}$$

If the case of small rotational displacements is assumed, the relationship between the time derivative of Euler angles and the angular velocity (2.20)— see also (3.19) in the discussion below— leads to

$$\boldsymbol{\mu}_E^{\mathrm{T}} \Delta\boldsymbol{\omega}_{dc} = \Delta\boldsymbol{\varphi}_{dc}^{\mathrm{T}} \boldsymbol{K}_o \boldsymbol{T}^{-1}(\boldsymbol{\varphi}_c) \Delta\boldsymbol{\omega}_{dc} \,, \tag{3.16}$$

which, in the absence of representation singularities, yields

$$\boldsymbol{T}^{\mathrm{T}}(\boldsymbol{\varphi}_c)\boldsymbol{\mu}_E = \boldsymbol{K}_o \Delta\boldsymbol{\varphi}_{dc} \,. \tag{3.17}$$

As a consequence, differently from (3.9), the dynamic behaviour for the rotational part is not uniquely determined by the choice of the impedance parameters but it does also depend on the orientation of the compliant frame with respect to the base frame through the matrix $\boldsymbol{T}^{\mathrm{T}}(\boldsymbol{\varphi}_c)$. Moreover, Equation (3.12) becomes ill-defined in the neighbourhood of a representation singularity; in particular, at such a singularity, moment components in the null space of $\boldsymbol{T}^{\mathrm{T}}$ do not generate any contribution to the dynamics of the orientation displacement, leading to a possible build-up of large values of contact moment.

The effect of the rotational stiffness can be better understood by considering an infinitesimal orientation displacement between Σ_d and Σ_c. From (3.12) or (3.17), in the absence of representation singularities, the elastic moment is

$$\boldsymbol{\mu}_E = \boldsymbol{T}^{\mathrm{T}}(\boldsymbol{\varphi}_c)\boldsymbol{K}_o\Delta\boldsymbol{\varphi}_{dc} \,. \tag{3.18}$$

In the case of an infinitesimal orientation displacement about $\boldsymbol{\varphi}_c$, from (2.20) it follows

$$\begin{aligned} \mathrm{d}(\Delta\boldsymbol{\varphi}_{dc}) &= (\dot{\boldsymbol{\varphi}}_d - \dot{\boldsymbol{\varphi}}_c)\Big|_{\boldsymbol{\varphi}_d = \boldsymbol{\varphi}_c} \mathrm{d}t \\ &= \boldsymbol{T}^{-1}(\boldsymbol{\varphi}_c)\Delta\boldsymbol{\omega}_{dc}\mathrm{d}t \,, \end{aligned} \tag{3.19}$$

where $\Delta\boldsymbol{\omega}_{dc} = \boldsymbol{\omega}_d - \boldsymbol{\omega}_c$ is the relative angular velocity between the two frames. Folding (3.19) into (3.18) written for an infinitesimal displacement $\mathrm{d}(\Delta\boldsymbol{\varphi}_{dc})$ gives

$$\boldsymbol{\mu}_E = \boldsymbol{T}^{\mathrm{T}}(\boldsymbol{\varphi}_c)\boldsymbol{K}_o\boldsymbol{T}^{-1}(\boldsymbol{\varphi}_c)\Delta\boldsymbol{\omega}_{dc}\mathrm{d}t \,. \tag{3.20}$$

Equation (3.20) reveals that the relationship between the orientation displacement and the elastic moment depends on the orientation of Σ_c. It follows that the property of task geometric consistency of the elastic force (3.11) is lost when considering the elastic moment (3.18), that is, the eigenvectors of the matrix $\boldsymbol{K}_o$ do not represent the three principal axes for the rotational stiffness.

3.2.3 Rotational impedance based on alternative Euler angles

The drawbacks discussed above can be mitigated by adopting the *alternative Euler angles orientation displacement* $\boldsymbol{\varphi}_{dc}$ that can be extracted from the rotation matrix

$${}^c\boldsymbol{R}_d = \boldsymbol{R}_c^{\mathrm{T}}\boldsymbol{R}_d. \tag{3.21}$$

Then, the *rotational impedance equation* at the end effector can be stated as

$$\boldsymbol{M}_o\ddot{\boldsymbol{\varphi}}_{dc} + \boldsymbol{D}_o\dot{\boldsymbol{\varphi}}_{dc} + \boldsymbol{K}_o\boldsymbol{\varphi}_{dc} = \boldsymbol{T}^{\mathrm{T}}(\boldsymbol{\varphi}_{dc}){}^c\boldsymbol{\mu} \,, \tag{3.22}$$

where $\boldsymbol{M}_o$, $\boldsymbol{D}_o$ and $\boldsymbol{K}_o$ are defined in a similar way to (3.12) and ${}^c\boldsymbol{\mu}$ is referred to Σ_c. This form of the rotational impedance equation was originally proposed in [13].

An advantage with respect to (3.12) is that now the impedance behaviour for the rotational part depends only on the relative orientation between Σ_d and Σ_c through the matrix $\boldsymbol{T}^{\mathrm{T}}(\boldsymbol{\varphi}_{dc})$. Hence, if XYZ Euler angles are adopted, representation singularities have a mitigated effect since they occur when $\beta_{dc} = \pm\pi/2$, i.e. for large end-effector orientation displacements.

From (3.22) the elastic moment is

$$ {}^{c}\boldsymbol{\mu}_{E} = \boldsymbol{T}^{-\mathrm{T}}(\boldsymbol{\varphi}_{dc})\boldsymbol{K}_{o}\boldsymbol{\varphi}_{dc} \,. \tag{3.23} $$

The infinitesimal orientation displacement about $\boldsymbol{\varphi}_{dc} = \mathbf{0}$ is

$$ \mathrm{d}\boldsymbol{\varphi}_{dc} = \dot{\boldsymbol{\varphi}}_{dc}\Big|_{\boldsymbol{\varphi}_{dc}=\mathbf{0}} \mathrm{d}t = \boldsymbol{T}^{-1}(\mathbf{0})\Delta^{c}\boldsymbol{\omega}_{dc}\mathrm{d}t \,. \tag{3.24} $$

Folding (3.24) into (3.23) written for an infinitesimal displacement $\mathrm{d}\boldsymbol{\varphi}_{dc}$ gives

$$ \begin{aligned} {}^{c}\boldsymbol{\mu}_{E} &= \boldsymbol{T}^{-\mathrm{T}}(\mathrm{d}\boldsymbol{\varphi}_{dc})\boldsymbol{K}_{o}\boldsymbol{T}^{-1}(\mathbf{0})\Delta^{c}\boldsymbol{\omega}_{dc}\mathrm{d}t \\ &\simeq \boldsymbol{T}^{-\mathrm{T}}(\mathbf{0})\boldsymbol{K}_{o}\boldsymbol{T}^{-1}(\mathbf{0})\Delta^{c}\boldsymbol{\omega}_{dc}\mathrm{d}t \,, \end{aligned} \tag{3.25} $$

where the first-order approximation $\boldsymbol{T}^{-\mathrm{T}}(\mathrm{d}\boldsymbol{\varphi}_{dc})\mathrm{d}t \simeq \boldsymbol{T}^{-\mathrm{T}}(\mathbf{0})\mathrm{d}t$ has been made. Equation (3.25) reveals that the relationship between the orientation displacement and the elastic moment is independent of the orientation of Σ_c. Notice, however, that the choice of Euler angles affects the resulting stiffness through the matrix $\boldsymbol{T}(\mathbf{0})$ which must be invertible; as already emphasized in Subsection 2.2.2 of the previous chapter, the widely-adopted ZYZ representation of Euler angles cannot be used here, being singular right at $\boldsymbol{\varphi}_{dc} = \mathbf{0}$. It is convenient, instead, to adopt the XYZ representation which gives $\boldsymbol{T}(\mathbf{0}) = \boldsymbol{I}$ and thus, for an infinitesimal displacement,

$$ {}^{c}\boldsymbol{\mu}_{E} \simeq \boldsymbol{K}_{o}\Delta^{c}\boldsymbol{\omega}_{dc}\mathrm{d}t \,. \tag{3.26} $$

As regards the property of task geometric consistency for the elastic moment (3.23), when $\boldsymbol{K}_o$ is a diagonal matrix and the XYZ representation of Euler angles is adopted, the i-th eigenvector $\boldsymbol{u}_{oi}$ of $\boldsymbol{K}_o = \mathrm{diag}\{\gamma_{o1}, \gamma_{o2}, \gamma_{o3}\}$ is the i-th column of the identity matrix. Hence, the orientation displacement of an angle ϑ_{dc} about $\boldsymbol{u}_{oi}$ is described by

$$ \boldsymbol{\varphi}_{dc} = \vartheta_{dc}\boldsymbol{u}_{oi} \,, \tag{3.27} $$

which, in view of the expression of $\boldsymbol{T}(\boldsymbol{\varphi}_{dc})$ for XYZ Euler angles in (2.34), leads to

$$ {}^{c}\boldsymbol{\mu}_{E} = \gamma_{oi}\vartheta_{dc}\boldsymbol{u}_{oi} \,, \tag{3.28} $$

representing an elastic moment about the same $\boldsymbol{u}_{oi}$ axis; thus, the vectors $\boldsymbol{u}_{oi}$ have the meaning of rotational stiffness principal axes. It can be easily recognized that the same property does not hold in general for a nondiagonal $\boldsymbol{K}_o$.

3.3 Impedance Equation in the Task Space

The analysis carried out in the previous section led to discover the major drawbacks of the operational space approach.

First of all, the occurrence of representation singularities, due to the use of minimal representation of orientation, in the interaction control is more critical than in the motion control. In fact, if in the latter it would be possible to plan a desired trajectory avoiding representation singularities, in the former end-effector trajectory is not a priori known, because it depends on the contact force and moment. By adopting the alternative Euler angles representation, a mitigation of this drawback has been achieved, but the problem can be completely overcome only by resorting to non-minimal representation of orientation displacements.

Moreover, the analysis of the impedance equation based on the task geometry has revealed that the adoption of the operational space framework does not allow, in general, the user to specify the impedance parameters in a way which is consistent with the task geometry, except for some special cases. As already discussed in the previous chapter about the desired trajectory specification in a consistent way with the task geometry, the only way to save the property of task geometric consistency is to adopt geometrical meaningful representations of end-effector orientation.

Since, as shown in Section 3.2, for the translational part of the impedance equation in the operational space approach, the property of task geometric consistency holds, in the following, only the rotational part of the impedance equation will be reformulated according to the *task space framework* [16].

3.3.1 Rotational impedance based on angle/axis

A class of geometrically meaningful representations of the mutual orientation between Σ_d and Σ_c can be given in terms of the *angle/axis orientation displacement*,

$$ {}^{c}\boldsymbol{o}_{dc} = f(\vartheta_{dc})\,{}^{c}\boldsymbol{r}_{dc} \,, \tag{3.29}$$

where ϑ_{dc} and ${}^{c}\boldsymbol{r}_{dc}$ correspond to ${}^{c}\boldsymbol{R}_{d}$, and $f(\vartheta_{dc})$ is any of the functions listed in Tab. 2.1. Those are strictly increasing smooth functions in an interval $(-\vartheta_M, \vartheta_M)$ with $\vartheta_M > 0$. Hence, the derivative $f'(\vartheta_{dc})$ of f with respect to ϑ_{dc} is strictly positive in that interval.

From (A.34) the angular velocity of Σ_d relative to Σ_c is given by

$$ \Delta^{c}\boldsymbol{\omega}_{dc} = {}^{c}\boldsymbol{\omega}_{d} - {}^{c}\boldsymbol{\omega}_{c} = \boldsymbol{R}_{c}^{\mathrm{T}}(\boldsymbol{\omega}_{d} - \boldsymbol{\omega}_{c}) \,. \tag{3.30}$$

The relationship between the time derivative of the angle/axis parameters and the relative angular velocity can be computed by taking into account the Rodrigues' formula (A.9) written for the rotation matrix ${}^{c}\boldsymbol{R}_{d}$

$$ \dot{\vartheta}_{dc} = {}^{c}\boldsymbol{r}_{dc}^{\mathrm{T}}\,{}^{c}\Delta\boldsymbol{\omega}_{dc} \tag{3.31a}$$

$$ {}^{c}\dot{\boldsymbol{r}}_{dc} = \frac{1}{2}\left((\boldsymbol{I} - {}^{c}\boldsymbol{r}_{dc}{}^{c}\boldsymbol{r}_{dc}^{\mathrm{T}})\cot\frac{\vartheta}{2} - \boldsymbol{S}({}^{c}\boldsymbol{r}_{dc})\right){}^{c}\Delta\boldsymbol{\omega}_{dc} \,. \tag{3.31b}$$

Differentiating (3.29), and using (3.31a),(3.31b), gives

$$ {}^{c}\dot{\boldsymbol{o}}_{dc} = \boldsymbol{\Omega}({}^{c}\boldsymbol{r}_{dc}, \vartheta_{dc})\Delta^{c}\boldsymbol{\omega}_{dc} \,, \tag{3.32}$$

where

$$\boldsymbol{\Omega} = \boldsymbol{\Omega}_{\|} + \boldsymbol{\Omega}_{\perp} \,, \tag{3.33}$$

with

$$\boldsymbol{\Omega}_{\|} = f'(\vartheta_{dc})^c\boldsymbol{r}_{dc}{}^c\boldsymbol{r}_{dc}^{\mathrm{T}} \tag{3.34a}$$

$$\boldsymbol{\Omega}_{\perp} = \frac{1}{2}f(\vartheta_{dc})\Big(\cot(\vartheta_{dc}/2)\big(\boldsymbol{I} - {}^c\boldsymbol{r}_{dc}{}^c\boldsymbol{r}_{dc}^{\mathrm{T}}\big) - \boldsymbol{S}({}^c\boldsymbol{r}_{dc})\Big) \,. \tag{3.34b}$$

The matrix $\boldsymbol{\Omega}_{\|}$ ($\boldsymbol{\Omega}_{\perp}$) projects the relative angular velocity $\Delta^c\boldsymbol{\omega}_{dc}$ in a direction parallel (orthogonal) to ${}^c\boldsymbol{o}_{dc}$. Also, notice that the following property of $\boldsymbol{\Omega}$ holds

$$\boldsymbol{\Omega}({}^c\boldsymbol{r}_{dc}, 0) = f'(0)\boldsymbol{I} \,, \tag{3.35}$$

which will be useful in the following.

In order to derive the impedance equation for the rotational part, since all the quantities now have a geometrical and physical meaning, it is possible to pursue the same energy-based argument adopted to derive the translational impedance equation. Let

$$\mathcal{T}_o = \frac{1}{2}\Delta^c\boldsymbol{\omega}_{dc}^{\mathrm{T}}\boldsymbol{M}_o\Delta^c\boldsymbol{\omega}_{dc} \tag{3.36}$$

express the rotational kinetic energy of a rigid body with inertia tensor $\boldsymbol{M}_o$ (symmetric and positive definite) and angular velocity $\Delta^c\boldsymbol{\omega}_{dc}$. It is worth pointing out that, from a rigourous physical viewpoint, $\mathcal{T}_o$ is representative of a pseudo-kinetic energy since it is defined in terms of a relative velocity. Nonetheless, it should be clear that, if the orientation of Σ_d is constant, then it would attain the meaning of a true kinetic energy.

Then, consider the potential energy

$$\mathcal{U}_o = \psi\,{}^c\boldsymbol{o}_{dc}^{\mathrm{T}}\boldsymbol{K}_o{}^c\boldsymbol{o}_{dc} \,, \tag{3.37}$$

where ψ is a positive constant depending on the particular choice of $f(\cdot)$, and $\boldsymbol{K}_o$ is a symmetric positive definite matrix.

Once the various energy contributions have been defined, the terms in the rotational impedance equation can be derived by considering the associated powers.

Computing the time derivative of (3.36) yields

$$\dot{\mathcal{T}}_o = {}^c\boldsymbol{\mu}_I^{\mathrm{T}}\Delta^c\boldsymbol{\omega}_{dc} \,, \tag{3.38}$$

where

$${}^c\boldsymbol{\mu}_I = \boldsymbol{M}_o\Delta^c\dot{\boldsymbol{\omega}}_{dc} \tag{3.39}$$

is the inertial moment and $\Delta^c\dot{\boldsymbol{\omega}}_{dc}$ denotes the time derivative of $\Delta^c\boldsymbol{\omega}_{dc}$ in (3.30). Further, taking the time derivative of (3.37) and accounting for (3.32) yields

$$\dot{\mathcal{U}}_o = {}^c\boldsymbol{\mu}_E^{\mathrm{T}}\Delta^c\boldsymbol{\omega}_{dc} \,, \tag{3.40}$$

where

$$ {}^c\boldsymbol{\mu}_E = 2\psi \boldsymbol{\Omega}^{\mathrm{T}}({}^c\boldsymbol{r}_{dc}, \vartheta_{dc}) \boldsymbol{K}_o {}^c\boldsymbol{o}_{dc} \tag{3.41} $$

is the elastic moment.

Finally, a dissipative contribution can be added as

$$ {}^c\boldsymbol{\mu}_D = \boldsymbol{D}_o \Delta^c\boldsymbol{\omega}_{dc} \,, \tag{3.42} $$

where $\boldsymbol{D}_o$ is a positive definite matrix characterizing a rotational damping at the end effector.

Therefore, the *rotational impedance equation* in the task space can be defined by adding the contributions (3.39),(3.42) and (3.41), i.e.

$$ \boldsymbol{M}_o \Delta^c\dot{\boldsymbol{\omega}}_{dc} + \boldsymbol{D}_o \Delta^c\boldsymbol{\omega}_{dc} + \boldsymbol{K}'_o {}^c\boldsymbol{o}_{dc} = {}^c\boldsymbol{\mu} \,, \tag{3.43} $$

where the equality ${}^c\boldsymbol{\mu} = {}^c\boldsymbol{\mu}_I + {}^c\boldsymbol{\mu}_D + {}^c\boldsymbol{\mu}_E$ has been imposed, and

$$ \boldsymbol{K}'_o = 2\psi \boldsymbol{\Omega}^{\mathrm{T}}({}^c\boldsymbol{r}_{dc}, \vartheta_{dc}) \boldsymbol{K}_o \,. \tag{3.44} $$

A similar energy-based argument has been adopted in [34] to define the so-called *spatial impedance*.

Notice that the rotational part of the impedance equation has been derived in terms of quantities all referred to Σ_c; this allows the impedance behaviour to be effectively expressed in terms of the relative orientation between Σ_d and Σ_c, no matter what the absolute orientation of the compliant frame with respect to the base frame is.

It is worth remarking that, by adopting an angle/axis representation of the orientation and pursuing an energy-based argument, the contributions in the rotational impedance equation correspond to physically meaningful energy terms; also, the velocity used is dual to the moment $\boldsymbol{\mu}$ exerted by the end effector, i.e. with no need of a transformation matrix depending on the actual end-effector orientation.

In the following, the analysis for small orientation displacements is carried out and consistency with the task geometry is investigated.

Consider an infinitesimal orientation displacement expressed as

$$ \mathrm{d}^c\boldsymbol{o}_{dc} = {}^c\dot{\boldsymbol{o}}_{dc} \Big|_{\vartheta_{dc}=0} \mathrm{d}t = \boldsymbol{\Omega}({}^c\boldsymbol{r}_{dc}, 0) \Delta^c\boldsymbol{\omega}_{dc}\mathrm{d}t = f'(0) \Delta^c\boldsymbol{\omega}_{dc}\mathrm{d}t \,, \tag{3.45} $$

where the property (3.35) has been exploited. Folding (3.45) into (3.41), written for an infinitesimal displacement about $\vartheta_{dc} = 0$, gives

$$ \begin{aligned} {}^c\boldsymbol{\mu}_E &= 2\psi \boldsymbol{\Omega}^{\mathrm{T}}({}^c\boldsymbol{r}_{dc}, \mathrm{d}\vartheta_{dc}) \boldsymbol{K}_o \mathrm{d}^c\boldsymbol{o}_{dc} \simeq 2\psi \left(f'(0)\right)^2 \boldsymbol{K}_o \Delta^c\boldsymbol{\omega}_{dc}\mathrm{d}t \\ &= \boldsymbol{K}_o \Delta^c\boldsymbol{\omega}_{dc}\mathrm{d}t \,, \end{aligned} \tag{3.46} $$

where the first-order approximation $\boldsymbol{\Omega}({}^c\boldsymbol{r}_{dc}, \mathrm{d}\vartheta_{dc}) \simeq f'(0)\boldsymbol{I}$ has been considered and the choice $\psi = 1/2(f'(0))^2$ has been made. Equation (3.46) clearly shows how the relationship between the orientation displacement and the elastic moment is independent of the orientation of Σ_c, and the problem of representation singularities is not of concern since $f'(0)$ is finite.

Concerning the property of task geometric consistency, the rotational stiffness matrix in the task space in (3.41) can be decomposed as

$$\boldsymbol{K}_o = \boldsymbol{U}_o \boldsymbol{\Gamma}_o \boldsymbol{U}_o^{\mathrm{T}} , \tag{3.47}$$

where $\boldsymbol{\Gamma}_o = \mathrm{diag}\{\gamma_{o1}, \gamma_{o2}, \gamma_{o3}\}$ and $\boldsymbol{U}_o = [\, \boldsymbol{u}_{o1} \quad \boldsymbol{u}_{o2} \quad \boldsymbol{u}_{o3} \,]$ are the eigenvalue matrix and the (orthogonal) eigenvector matrix, respectively. Then, considering an orientation displacement by an angle ϑ_{dc} about the i-th eigenvector

$$\boldsymbol{o}_{dc}^{c} = f(\vartheta_{dc})\boldsymbol{u}_{oi} , \tag{3.48}$$

and taking into account the decomposition of $\boldsymbol{\Omega}$ into the two terms (3.34a), (3.34b) yields

$$^{c}\boldsymbol{\mu}_E = 2\psi \boldsymbol{\Omega}_{\|}^{\mathrm{T}}(\boldsymbol{u}_{oi}, \vartheta_{dc})\gamma_{oi} f(\vartheta_{dc})\boldsymbol{u}_{oi} = 2\psi f'(\vartheta_{dc})\gamma_{oi} f(\vartheta_{dc})\boldsymbol{u}_{oi} \,. \tag{3.49}$$

This represents an elastic moment about the same $\boldsymbol{u}_{oi}$ axis which is in the same direction of the orientation displacement since $f'(\vartheta_{dc}) > 0$. Therefore, the rotational stiffness matrix can be expressed in terms of three parameters γ_{oi} representing the stiffness about three principal axes $\boldsymbol{u}_{oi}$, i.e. in a consistent way with the task geometry. A generalization to the case of mutual interaction between rotational and translational stiffness can be found in [12].

3.3.2 Rotational impedance based on quaternion

With reference to the different angle/axis representations of orientation displacement in Table 2.1, a special case is constituted by the *quaternion orientation displacement*. Such a representation has the advantage over other angle/axis representations to avoid representation singularities and it can completely substitute the rotation matrices, in the sense already explained in Subsection 2.2.4. In fact, the orientation displacement $^{c}\boldsymbol{o}_{dc}$ with $f(\vartheta_{dc}) = \sin(\vartheta_{dc}/2)$ (the choice corresponding to the unit quaternion) is the vector part of the quaternion

$$\mathcal{Q}_{dc} = \{\eta_{dc}, {}^{c}\boldsymbol{\epsilon}_{dc}\} = \mathcal{Q}_c^{-1} * \mathcal{Q}_d \,. \tag{3.50}$$

expressing the mutual orientation between Σ_d and Σ_c.

With this choice, in view of (3.37), the expression of the potential energy becomes

$$\mathcal{U}_o = 2\,{}^{c}\boldsymbol{\epsilon}_{dc}^{\mathrm{T}} \boldsymbol{K}_o {}^{c}\boldsymbol{\epsilon}_{dc} , \tag{3.51}$$

where it has been set $\psi = 2$. Even though the potential energy is expressed in terms only of the vector part of the quaternion, it can be shown that $\mathcal{U}_o$ coincides with the rotational elastic energy associated with a torsional spring with rotational stiffness matrix $\boldsymbol{K}_o$ acting so as to align Σ_c with Σ_d.

In view of (3.43), the resulting impedance equation for the rotational part becomes

$$\boldsymbol{M}_o \Delta^{c}\dot{\boldsymbol{\omega}}_{dc} + \boldsymbol{D}_o \Delta^{c}\boldsymbol{\omega}_{dc} + \boldsymbol{K}_o'\,{}^{c}\boldsymbol{\epsilon}_{dc} = {}^{c}\boldsymbol{\mu} , \tag{3.52}$$

where

$$\boldsymbol{K}_o' = 2\boldsymbol{E}^{\mathrm{T}}(\eta_{dc}, {}^c\boldsymbol{\epsilon}_{dc})\boldsymbol{K}_o \, , \tag{3.53}$$

with $\boldsymbol{E}$ as in (A.29).

In the case of free motion, it is worth finding the equilibria of the rotational impedance equation (3.52). These should occur whenever Σ_d and Σ_c are aligned.

Consider the Hamiltonian contribution [60]

$$\mathcal{H}_o = \mathcal{T}_o + \mathcal{U}_o \, , \tag{3.54}$$

associated with the rotational motion, which is a positive definite function. Taking the time derivative of (3.54) and accounting for (3.36),(3.37) along with (3.52), (3.53) yields

$$\dot{\mathcal{H}}_o = -\Delta{}^c\boldsymbol{\omega}_{dc}^{\mathrm{T}}\boldsymbol{D}_o\Delta{}^c\boldsymbol{\omega}_{dc} + {}^c\boldsymbol{\mu}^{\mathrm{T}}\Delta{}^c\boldsymbol{\omega}_{dc} \, . \tag{3.55}$$

If ${}^c\boldsymbol{\mu} = \mathbf{0}$, $\dot{\mathcal{H}}_o$ in (3.55) vanishes if and only if $\Delta{}^c\boldsymbol{\omega}_{dc} = \mathbf{0}$; hence from (3.52) it follows that ${}^c\boldsymbol{\epsilon}_{dc}$ asymptotically tends to the invariant set described by

$${}^c\boldsymbol{\mu}_E = 2\left(\eta_{dc}\boldsymbol{K}_o{}^c\boldsymbol{\epsilon}_{dc} + \boldsymbol{S}({}^c\boldsymbol{\epsilon}_{dc})\boldsymbol{K}_o{}^c\boldsymbol{\epsilon}_{dc}\right) = \mathbf{0} \, , \tag{3.56}$$

where (3.53) has been exploited.

By observing that the two terms in (3.56) are mutually orthogonal, the following sets of equilibria are found:

$$\mathcal{E}_1 = \{\eta_{dc} = \pm 1, \quad {}^c\boldsymbol{\epsilon}_{dc} = \mathbf{0}, \quad \Delta{}^c\boldsymbol{\omega}_{dc} = \mathbf{0}\} \tag{3.57a}$$

$$\mathcal{E}_2 = \{\eta_{dc} = 0, {}^c\boldsymbol{\epsilon}_{dc} \colon \boldsymbol{K}_o{}^c\boldsymbol{\epsilon}_{dc} = \gamma_{oi}{}^c\boldsymbol{\epsilon}_{dc}, \|{}^c\boldsymbol{\epsilon}_{dc}\| = 1, \Delta{}^c\boldsymbol{\omega}_{dc} = \mathbf{0}\} \, , \tag{3.57b}$$

where $\gamma_{oi} > 0$ denotes an eigenvalue of matrix $\boldsymbol{K}_o$.

The equilibria in $\mathcal{E}_2$ are unstable. To see this, consider the Hamiltonian contribution (3.54) which, in view of (3.55), is a decreasing function. At any of the equilibria in (3.57b), it is

$$\mathcal{H}_{o,\infty} = 2\gamma_{oi}{}^c\boldsymbol{\epsilon}_{dc}^{\mathrm{T}}{}^c\boldsymbol{\epsilon}_{dc} = 2\gamma_{oi} \tag{3.58}$$

where (A.15) has been used. Consider a small perturbation around the equilibrium with $\eta_{dc} = \sigma$, ${}^c\boldsymbol{\epsilon}_{dc}$ such that ${}^c\boldsymbol{\epsilon}_{dc}^{\mathrm{T}}{}^c\boldsymbol{\epsilon}_{dc} = 1 - \sigma^2$, $\Delta{}^c\boldsymbol{\omega}_{dc} = \mathbf{0}$ and $\boldsymbol{K}_o{}^c\boldsymbol{\epsilon}_{dc} = \gamma_{oi}{}^c\boldsymbol{\epsilon}_{dc}$. The perturbed Hamiltonian contribution is

$$\mathcal{H}_{o,\sigma} = 2\gamma_{oi}(1 - \sigma^2) < \mathcal{H}_{o,\infty} \tag{3.59}$$

and thus, since (3.54) is decreasing, $\mathcal{H}_o$ will never return to $\mathcal{H}_{o,\infty}$, implying that those equilibria are all unstable. Notice that, at such equilibria, Σ_d is anti-aligned with Σ_c with respect to the axis of the mutual rotation ${}^c\boldsymbol{R}_d$ between the two frames.

It can be concluded that ${}^c\boldsymbol{\epsilon}_{dc}$ must converge to $\mathcal{E}_1$. Interestingly enough, the two equilibria in $\mathcal{E}_1$ both give the same mutual orientation ${}^c\boldsymbol{R}_d = \boldsymbol{I}$, thus implying the alignment of Σ_d with Σ_c, so as wished.

3.4 Impedance Control Law

Once the impedance equation has been defined, a proper control law is devised to ensure an end-effector behaviour realizing the programmed mechanical impedance. It is a well-known concept that in industrial robots the problem of joint friction, and other different torque disturbances, greatly affect the performance of the impedance controller when this is implemented by designing the joint driving torques only on the basis of the impedance parameters [9]. A solution to this problem is to adopt an inner motion control loop in combination with an outer force loop [30]. The strategy is schematically depicted in the block scheme of Fig. 3.1.

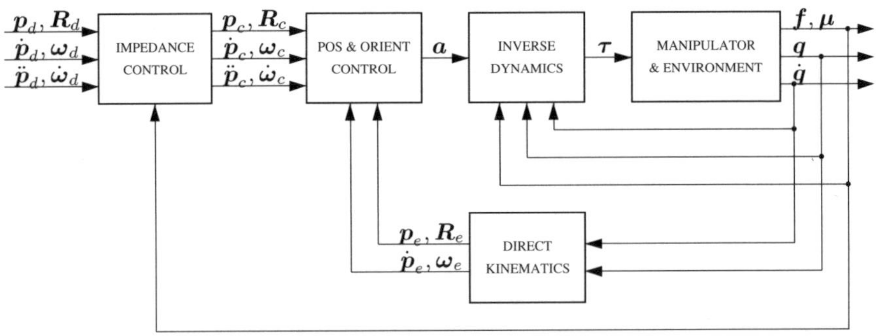

Fig. 3.1. Impedance control with inner motion control loop

In the impedance control, the compliant behaviour is imposed to the end-effector as long as tracking of position and orientation of the compliant frame Σ_c is guaranteed. To do this, it is possible to resort to the various motion control schemes described in Chapter 2, depending on the particular representation chosen for orientation displacements.

Therefore, with reference to the dynamic model of the robot in contact with a compliant environment in (2.6) and by adopting the inverse dynamics control technique (2.7), the joint driving torques can be taken as

$$\boldsymbol{\tau} = \boldsymbol{B}(\boldsymbol{q})\boldsymbol{J}^{-1}(\boldsymbol{q})\left(\boldsymbol{a} - \dot{\boldsymbol{J}}(\boldsymbol{q},\dot{\boldsymbol{q}})\right) + \boldsymbol{C}(\boldsymbol{q},\dot{\boldsymbol{q}})\dot{\boldsymbol{q}} + \boldsymbol{F}\dot{\boldsymbol{q}} + \boldsymbol{g}(\boldsymbol{q}) + \boldsymbol{J}(\boldsymbol{q})^{\mathrm{T}}\boldsymbol{h}\,, \tag{3.60}$$

where the contribution relative to the contact force and moment $\boldsymbol{h} = [\,\boldsymbol{f}^{\mathrm{T}} \;\; \boldsymbol{\mu}^{\mathrm{T}}\,]^{\mathrm{T}}$ can be recognized. It is clear that to implement such a control algorithm a force sensor is needed, not only to cancel the joint torques corresponding to the end-effector contact force, but also to compute the resolved acceleration $\boldsymbol{a} = [\,\boldsymbol{a}_p^{\mathrm{T}} \;\; \boldsymbol{a}_o^{\mathrm{T}}\,]^{\mathrm{T}}$. In fact, the compliant trajectory $\boldsymbol{p}_c(t), \boldsymbol{R}_c(t)$ to be tracked is computed by forward integration of the impedance equation with input the measured force and moment.

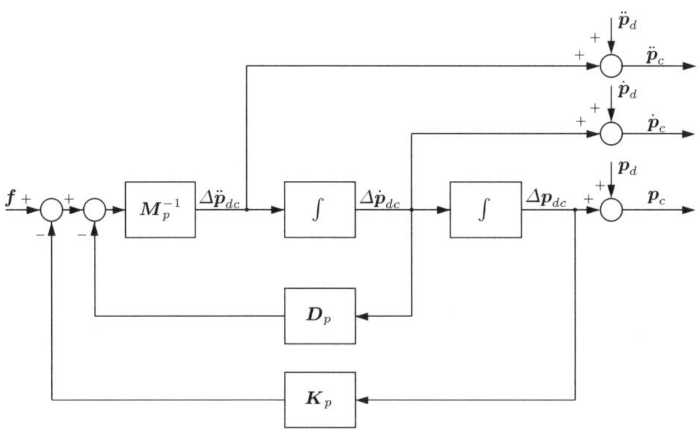

Fig. 3.2. Translational impedance equation

3.4.1 Operational space control

First, the translational impedance equation (3.9) is integrated (see Fig. 3.2), with input $\boldsymbol{f}$ and the reference trajectory $\boldsymbol{p}_c(t), \dot{\boldsymbol{p}}_c(t), \ddot{\boldsymbol{p}}_c(t)$ is input to the inner motion control law (3.60), to compute the resolved linear acceleration

$$\boldsymbol{a}_p = \ddot{\boldsymbol{p}}_c + \boldsymbol{K}_{Dp}\Delta\dot{\boldsymbol{p}}_{ce} + \boldsymbol{K}_{Pp}\Delta\boldsymbol{p}_{ce} \, , \tag{3.61}$$

where

$$\Delta\boldsymbol{p}_{ce} = \boldsymbol{p}_c - \boldsymbol{p}_e \tag{3.62}$$

is the position tracking error between Σ_c and Σ_e and $\boldsymbol{K}_{Dp}, \boldsymbol{K}_{Pp}$ are positive definite matrix gains. Notice that owing to the use of the inner motion control loop, the tracking performance of the control system depends only on the gains of the inner loop and not on the impedance parameters, and thus the former can be chosen to reject torque disturbances, e.g., joint friction, and obtain a good dynamic performance. On the other hand, the latter can be chosen only on the basis of the interaction task, i.e., to ensure a proper compliant behaviour of the robot end effector.

Concerning the rotational part of the impedance control, the resolved angular acceleration $\boldsymbol{a}_o$ can be computed in two different forms. The first is obtained by adopting the classical Euler angles representation of end-effector orientation displacements; then the rotational impedance equation (3.12) is integrated in a way formally equal to that represented in Fig. 3.2, and the reference trajectory $\boldsymbol{\varphi}_c(t), \dot{\boldsymbol{\varphi}}_c(t), \ddot{\boldsymbol{\varphi}}_c(t)$ is input to the inner motion control law (3.60) to compute

$$\boldsymbol{a}_o = \boldsymbol{T}(\boldsymbol{\varphi}_e)\left(\ddot{\boldsymbol{\varphi}}_c + \boldsymbol{K}_{Do}\Delta\dot{\boldsymbol{\varphi}}_{ce} + \boldsymbol{K}_{Po}\Delta\boldsymbol{\varphi}_{ce}\right) + \dot{\boldsymbol{T}}(\boldsymbol{\varphi}_e, \dot{\boldsymbol{\varphi}}_e)\dot{\boldsymbol{\varphi}}_e \, , \tag{3.63}$$

where

$$\Delta\boldsymbol{\varphi}_{ce} = \boldsymbol{\varphi}_c - \boldsymbol{\varphi}_e \tag{3.64}$$

is the orientation tracking error between Σ_c and Σ_e.

The second form can be adopted to partially overcome the drawbacks of representation singularities and task geometric inconsistency. It is based on the alternative Euler angles representation (3.22). By integrating this equation, it is possible to compute the reference trajectory $\varphi_{dc}, \dot{\varphi}_{dc}, \ddot{\varphi}_{dc}$ and thus compute $\boldsymbol{a}_o$ as

$$\begin{aligned}\boldsymbol{a}_o = &\dot{\boldsymbol{\omega}}_d - \dot{\boldsymbol{T}}_e(\boldsymbol{\varphi}_{de}, \dot{\boldsymbol{\varphi}}_{de})\dot{\boldsymbol{\varphi}}_{de} \\ &- \boldsymbol{T}_e(\boldsymbol{\varphi}_{de})\left(\ddot{\boldsymbol{\varphi}}_{dc} + \boldsymbol{K}_{Do}(\dot{\boldsymbol{\varphi}}_{dc} - \dot{\boldsymbol{\varphi}}_{de}) + \boldsymbol{K}_{Po}(\boldsymbol{\varphi}_{dc} - \boldsymbol{\varphi}_{de})\right) ,\end{aligned} \tag{3.65}$$

where the orientation control acts as to take φ_{de} to coincide with φ_{dc} which ultimately implies that Σ_e is aligned with Σ_c.

3.4.2 Task space control

As already seen for the motion control problem, in the task space framework [16], the translational part of the controller can be equal to that of the operational space approach, while differences arise for the rotational part. Two characteristics distinguish the former framework from the latter, namely the task geometric consistency, ensured by the use of the angle/axis representation of orientation, and the absence of representation singularities when the unit quaternion is adopted.

With reference to the class of angle/axis representations, the rotational impedance equation (3.43) together with (3.32) can be integrated to obtain the references for an inner motion loop, so that the resolved angular acceleration can be computed, i.e.

$$\boldsymbol{a}_o = \dot{\boldsymbol{\omega}}_c + \boldsymbol{K}_{Do}\Delta\boldsymbol{\omega}_{ce} + \boldsymbol{K}_{Po}\boldsymbol{o}_{ce} , \tag{3.66}$$

where the expression of $\boldsymbol{o}_{ce}$ depends on the particular choice of $f(\vartheta)$ made among all the possible listed in Tab. 2.1.

If quaternion displacement is adopted, the previous equation takes the form

$$\boldsymbol{a}_o = \dot{\boldsymbol{\omega}}_c + \boldsymbol{K}_{Do}\Delta\boldsymbol{\omega}_{ce} + \boldsymbol{K}_{Po}\boldsymbol{\epsilon}_{ce} \tag{3.67}$$

$\boldsymbol{\epsilon}_{ce}$ being the vector part, referred to the base frame, of the quaternion

$$\mathcal{Q}_{ce} = \mathcal{Q}_e^{-1} * \mathcal{Q}_c \tag{3.68}$$

where $\mathcal{Q}_c, \boldsymbol{\omega}_c, \dot{\boldsymbol{\omega}}_c$ can be computed via forward integration of the impedance equation (3.52) together with the quaternion propagation (A.28a),(A.28b), as illustrated in Fig. 3.3.

3.4.3 Experiments

The above impedance control schemes have been tested in a number of experiments on the six-joint and the seven-joint industrial robots with open control architecture and force/torque sensor described in Section 1.4.

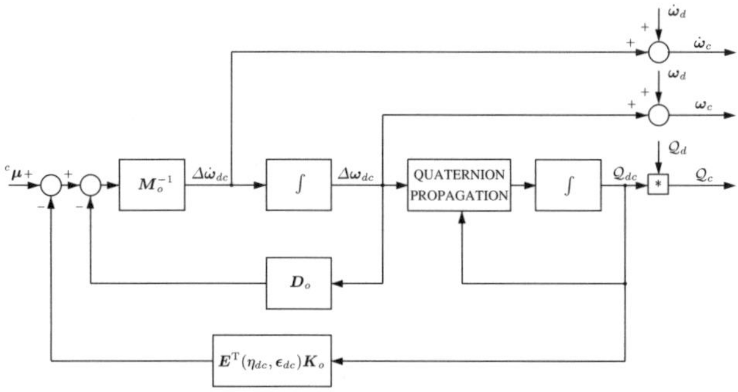

Fig. 3.3. Rotational impedance equation

An end effector has been built as a steel stick with a wooden disk of 5.5 cm radius at the tip (see Fig. 1.3). The end-effector frame has its origin at the center of the disk and its approach axis normal to the disk surface and pointing outwards.

The performance of the quaternion-based six-DOF task space impedance control has been compared to that of the two six-DOF operational space impedance control schemes based on Euler angles. An analysis of the computational burden for the three control schemes, based on (3.60),(2.11),(3.61) and (3.9), has been carried out for the available hardware, leading to a total time of: 0.264 ms for the impedance control using (3.63),(3.12), 0.230 ms for the impedance control using (3.65),(3.22), and 0.195 ms for the impedance control using (3.67),(3.52). Details on the computational load in terms of floating-point operations and transcendental functions number are given in Table 3.1.

Table 3.1. Computational load for the three six-DOF impedance control schemes.

Orientation error	*Flops*	*Funcs*
Classical Euler angles	1420	30
Alternative Euler angles	1467	24
Quaternion	1429	15

First case study: Interaction with the environment. The first case study has been developed to analyse the interaction with the environment. This is constituted by a flat plexiglas surface. The translational stiffness at the contact between the end effector and the surface is of the order of 10^4 N/m, while the rotational stiffness for small angles is of the order of 20 Nm/rad.

The task consists in taking the disk in contact with the surface at an angle of unknown magnitude (Fig. 3.4). The end-effector desired position is required to make a straight-line motion with a vertical displacement of $-0.24\,\mathrm{m}$ along the Z_b-axis of the base frame. The trajectory along the path is generated according to a fifth-order interpolating polynomial with null initial and final velocities and accelerations, and a duration of 7 s. The end-effector desired orientation is required to remain constant during the task. The surface is placed (horizontally) in the X_bY_b-plane in such a way as to obstruct the desired end-effector motion, both for the translational part and for the rotational part.

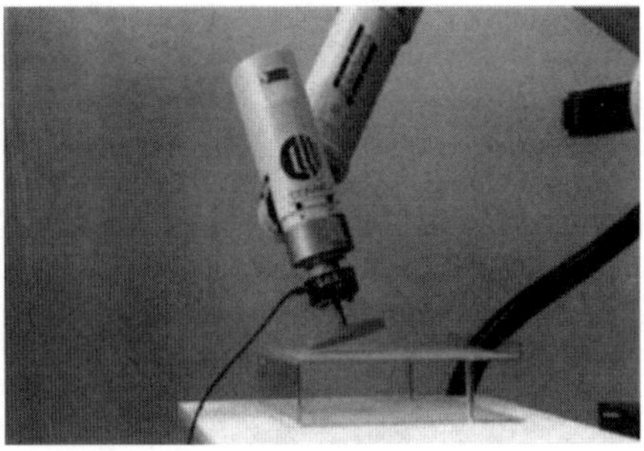

Fig. 3.4. End effector in contact with plexiglas surface

The parameters of the translational part of the six-DOF impedance equation (3.9) have been set to

$$\boldsymbol{M}_p = 9\boldsymbol{I}, \quad \boldsymbol{D}_p = 2000\boldsymbol{I}, \quad \boldsymbol{K}_p = 700\boldsymbol{I}\ ,$$

while the parameters of the rotational part of the six-DOF impedance equation (3.52) have been set to

$$\boldsymbol{M}_o = 0.4\boldsymbol{I}, \quad \boldsymbol{D}_o = 5\boldsymbol{I}, \quad \boldsymbol{K}_o = 2\boldsymbol{I}\ .$$

Notice that the stiffness matrices have been chosen so as to ensure a compliant behaviour at the end effector (limited values of contact force and moment) along all the directions of the Cartesian space, even though this is not strictly required by the task. This is the typical choice to be made when the geometry of the contact is scarcely known or it can change during the constrained motion. The damping matrices have been chosen so as to guarantee a well-damped behaviour.

The gains of the inner motion control loop actions in (3.61),(3.67) have been set to

$$\boldsymbol{K}_{Pp} = 2025\boldsymbol{I}, \quad \boldsymbol{K}_{Dp} = 65\boldsymbol{I}$$

for the translational part, and

$$\boldsymbol{K}_{Po} = 4500\boldsymbol{I}, \quad \boldsymbol{K}_{Do} = 65\boldsymbol{I}$$

for the rotational part.

The results in Fig. 3.5 show the effectiveness of the quaternion-based six-DOF impedance control. After the contact, the component of the position error between Σ_d and Σ_e $\Delta \boldsymbol{p}_{de} = \boldsymbol{p}_d - \boldsymbol{p}_e$ along the Z_b-axis significantly deviates from zero, as expected, while small errors can be seen also for the components along the X_b- and the Y_b-axis due to contact friction. As for the orientation error, all the components of the orientation displacement between Σ_d and Σ_e ${}^e\boldsymbol{\epsilon}_{de}$ significantly deviate from zero since the end-effector frame has to rotate with respect to the base frame after the contact in order to comply with the surface. Also, in view of the imposed task, a prevailing component of the contact force can be observed along the Z_b-axis after the contact, while the small components along the X_b- and the Y_b-axis arise as a consequence of the above end-effector deviation. As for the contact moment referred to Σ_d, the component about the Z_b-axis is small, as expected. It can be recognized that all the above quantities reach constant steady-state values after the desired motion is stopped. The oscillations on the force and moment during the transient can be mainly ascribed to slipping of the disk on the surface after the contact.

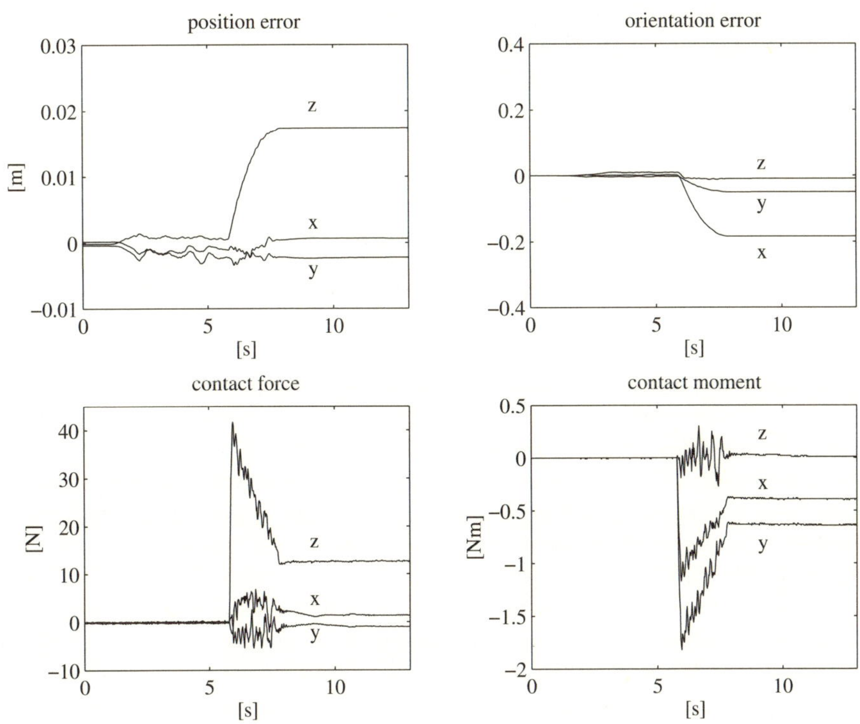

Fig. 3.5. Experimental results under six-DOF impedance control based on quaternion in the first case study

In sum, it can be asserted that a compliant behaviour is successfully achieved. A similar performance has been obtained also with the six-DOF impedance control schemes based on the Euler angles error, i.e. by using either (3.12) or (3.22) in lieu of (3.52). This fact can be explained because both the absolute end-effector orientation in (3.12) and the relative orientation in (3.22) keep far from representation singularities. The results are not reported here for brevity.

Second case study: Representation singularity. The second case study is aimed at testing the performance of the quaternion-based compared to the Euler angles-based six-DOF impedance control, when the end-effector orientation is close to a representation singularity of $\boldsymbol{T}$. The end effector and the surface are the same as in the previous case study.

The end-effector desired position is required to make a straight-line motion with a horizontal displacement of 0.085 m along the X_b-axis of the base frame. The trajectory along the path is generated according to a fifth-order interpolating polynomial with null initial and final velocities and accelerations, and a duration of 5 s. The end-effector desired orientation is required to remain constant during the task. The surface is now placed vertically in such a way as to obstruct the desired end-effector motion, only for the rotational part though. Therefore, no impedance control has been accomplished for the translational part, i.e. $\boldsymbol{p}_c$ in (3.61) coincides with $\boldsymbol{p}_d$.

The parameters of the quaternion-based impedance equations (3.9),(3.52) are set to

$$\boldsymbol{M}_p = 10\boldsymbol{I} \quad \boldsymbol{D}_p = 600\boldsymbol{I}, \quad \boldsymbol{K}_p = 1000\boldsymbol{I}\,,$$

for the translational part, and

$$\boldsymbol{M}_o = 0.25\boldsymbol{I}, \quad \boldsymbol{D}_o = 3.5\boldsymbol{I}, \quad \boldsymbol{K}_o = 2.5\boldsymbol{I}$$

for the rotational part. In order to carry out a comparison, the impedance control based on the Euler angles has also been tested. The parameters of the rotational impedance equation (3.12) have been set to the same values as for the quaternion. As regards the gains of the inner motion control loop, these have been chosen equal to those in the previous experiment for both types of impedance control schemes.

The results in Figs. 3.6 and 3.7 show the significant differences occurring in the performance of the two schemes. For the impedance control based on (3.12), large values of contact force and moment are generated since the rotational impedance equation suffers from ill-conditioning of the matrix $\boldsymbol{T}(\boldsymbol{\varphi}_c)$; this phenomenon is not present for the quaternion-based impedance control based on (3.52) since representation singularities are not involved in the rotational impedance equation. On the other hand, testing of the impedance control based on the alternative Euler angles in (3.22) has revealed a performance as good as the quaternion-based impedance control, since the orientation displacement $\boldsymbol{\varphi}_{dc}$ is kept far from a representation singularity. Hence, the results are not reported here for brevity.

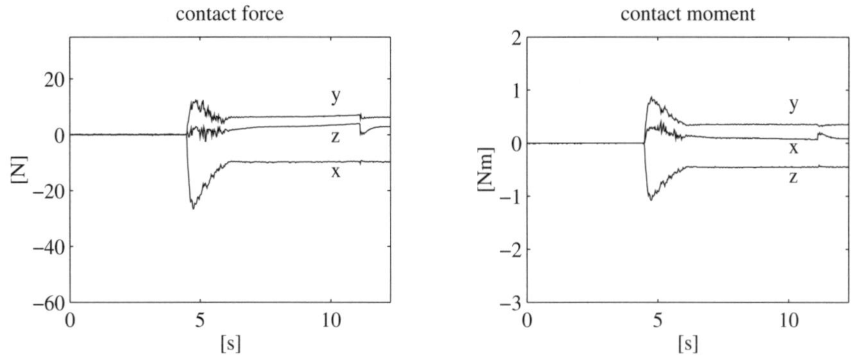

Fig. 3.6. Experimental results under six-DOF impedance control based on quaternion in the second case study

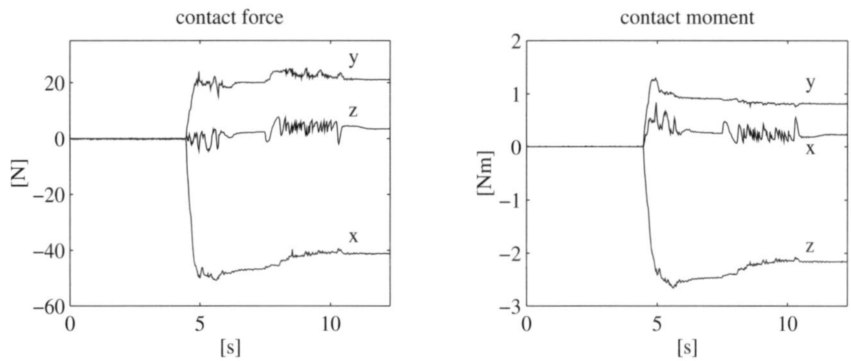

Fig. 3.7. Experimental results under six-DOF impedance control based on the classical Euler angles in the second case study

To sum up, it can be concluded that both the impedance control based on the alternative Euler angles and the quaternion-based impedance control perform better than the impedance control based on the classical Euler angles, as far as interaction with the environment is concerned.

Third case study: Task geometric consistency. Another case study has been developed to analyse the property of task geometric consistency when an external moment is applied at the end effector. The quaternion-based impedance control and the impedance control based on Euler angles have been tested.

The stiffness matrices of the rotational part of the impedance equations (3.52),(3.12) have been taken as diagonal matrices; $\boldsymbol{K}_o$ has been chosen as in (3.47) with $\boldsymbol{U}_o = \boldsymbol{I}$ and $\boldsymbol{\Gamma}_o = 2.5\boldsymbol{I}$ for both schemes. The remaining parameters of the rotational impedance have been set to $\boldsymbol{M}_o = 0.25\boldsymbol{I}$ and $\boldsymbol{D}_o = 1.5\boldsymbol{I}$ for both schemes. No impedance control has been accomplished for the translational part. The gains of the inner motion control loop have been chosen equal to those in the previous case study.

The position and orientation of the desired frame are required to remain constant, and a torque is applied about the approach axis of Σ_d; the torque is taken from zero to 2.5 Nm according to a linear interpolating polynomial with 4th-order blends and a total duration of 1 s.

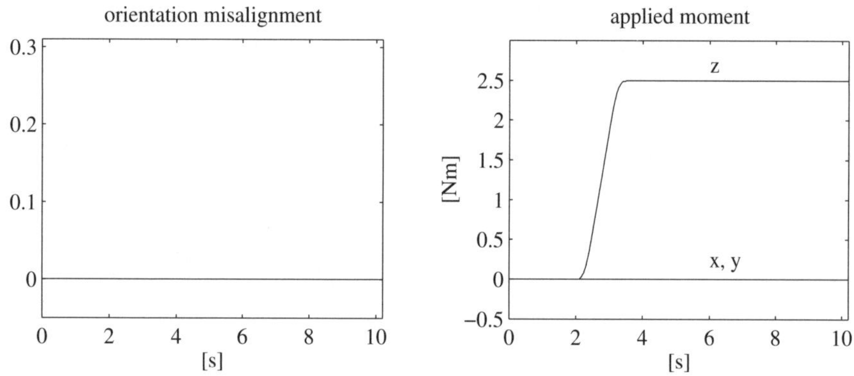

Fig. 3.8. Experimental results under six-DOF impedance control based on quaternion in the third case study

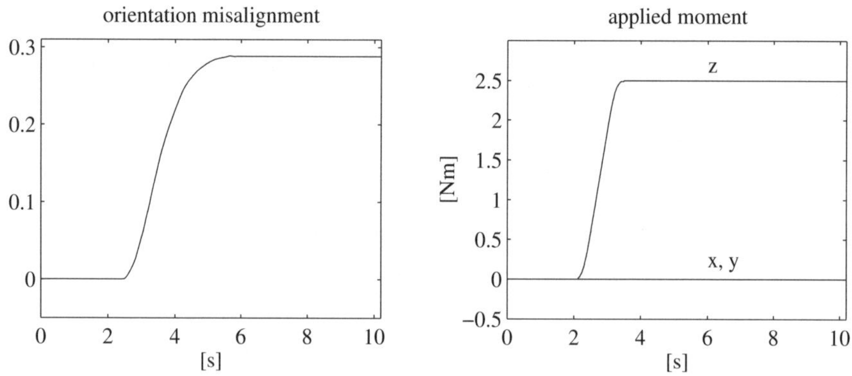

Fig. 3.9. Experimental results under six-DOF impedance control based on classical Euler angles in the third case study

The results in Figs. 3.8 and 3.9 show the different performance in terms of the orientation misalignment δ defined as the norm of the vector product between the orientation error and the unit vector $\boldsymbol{u}_{o3}$ of the approach axis of Σ_d, i.e.

$$\delta = \|\boldsymbol{S}({}^{e}\boldsymbol{\epsilon}_{de})\boldsymbol{u}_{o3}\| \,.$$

For the impedance control based on (3.12) the instantaneous axis of rotation of Σ_e changes, while remarkably no misalignment occurs for the impedance control based on (3.52). The impedance control based on (3.22) has also been tested and its performance is as good as that of the quaternion-based control; hence, the results are not reported for brevity.

Fourth case study: Non-diagonal rotational stiffness. In the fourth case study, the quaternion-based impedance control and the impedance control based on the alternative Euler angles have been tested when the rotational stiffness is chosen as a non-diagonal matrix. The impedance control based on the classical Euler angles has been ruled out in view of the poor results of the previous experiment.

The principal axis of the stiffness matrices of the rotational impedance equations (3.52),(3.22) are rotated with respect to the coordinate axes of Σ_d; $\boldsymbol{K}_o$ has been chosen as in (3.47) with

$$\boldsymbol{U}_o = \begin{bmatrix} 0.8047 & -0.3106 & 0.5059 \\ 0.5059 & 0.8047 & -0.3106 \\ -0.3106 & 0.5059 & 0.8047 \end{bmatrix} \qquad \boldsymbol{\Gamma}_o = \begin{bmatrix} 4 & 0 & 0 \\ 0 & 1 & 0 \\ 0 & 0 & 2.5 \end{bmatrix}$$

for both schemes. The remaining parameters of the rotational impedance have been set to $\boldsymbol{M}_o = 0.25\boldsymbol{I}$ and $\boldsymbol{D}_o = 1.5\boldsymbol{I}$ for both schemes. As above, no impedance control has been accomplished for the translational part, and the gains of the inner motion control loop have been chosen equal to those in the previous case study. A torque has been applied about the axis whose unit vector is $\boldsymbol{u}_{o3}$; the torque is taken from zero to $-1.5\,\mathrm{Nm}$ according to a linear interpolating polynomial with 4th-order blends and a total duration of 1 s.

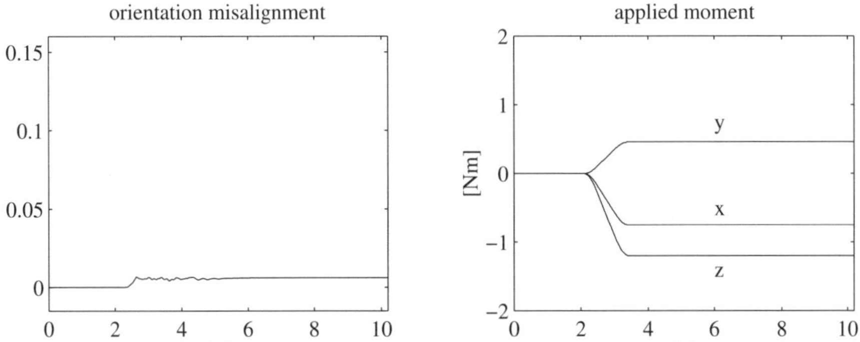

Fig. 3.10. Experimental results under six-DOF impedance control based on quaternion in the fourth case study

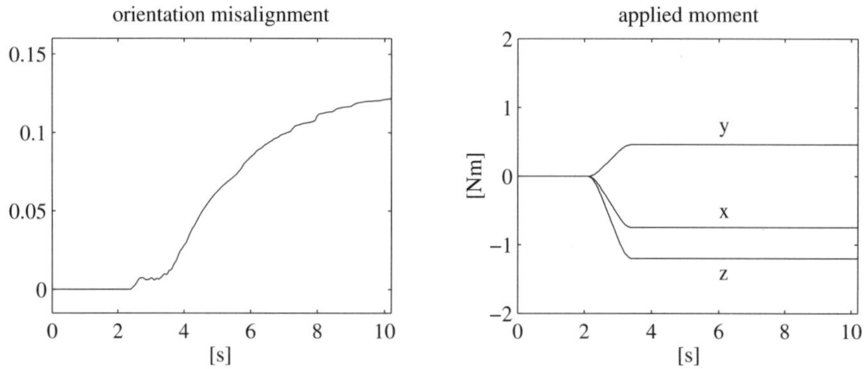

Fig. 3.11. Experimental results under six-DOF impedance control based on alternative Euler angles in the fourth case study

The results in Figs. 3.10 and 3.11 show the significant differences occurring in terms of the orientation misalignment δ. It can be seen that the instantaneous axis of rotation of Σ_e does not appreciably rotate with the impedance control based on (3.52), given the performance of the inner loop acting on the end-effector orientation error. Instead, a significant misalignment occurs with the impedance control based on (3.22).

To sum up, it can be concluded that the quaternion-based impedance control performs better than both impedance control schemes based on the Euler angles as far as task geometric consistency is concerned.

3.5 Redundant Manipulators

Up unto this point, the case of a non-redundant manipulator has been assumed, but, in general, a robot manipulator can have more *degrees-of-mobility* (its joints) than the task number of *degrees-of-freedom.* Since, in this work six-DOF interaction tasks are of interest, a manipulator is kinematically redundant if the number of its joints is greater than six.

In this case, there exists infinite joint motions that produce the same end-effector motion. In particular, even when the end effector is at rest, it is possible to generate an *internal motion* at the joints. As a minimal requirement, such motion should be made stable. Moreover, since the robot has to interact with the environment, a further requirement is necessary, i.e., joint acceleration in the null-space has to produce no end-effector force and vice-versa. In other words, the end-effector force corresponding to a given joint torque is required not to be affected by internal motions, as the contact dynamics should be influenced only by the impedance control loop. On the other hand, the internal motion could be keenly utilized to meet additional task requirements besides the execution of the end-effector trajectory, thus providing a *redundancy resolution*.

Redundancy can be solved either at a kinematic level, which is in the first stage of a kinematic control strategy, or at dynamic level by suitably modifying the inverse dynamics control law [41, 28]. The latter approach is pursued hereafter in order to meet all the above requirements.

Since the Jacobian matrix for a redundant manipulator has more columns than rows ($n > 6$), a suitable right inverse of $\boldsymbol{J}$ is to be used in lieu of $\boldsymbol{J}^{-1}$. Hence, in lieu of (2.11), the new control input in (2.7) can be chosen as

$$\boldsymbol{\phi} = \boldsymbol{J}^{\dagger}(\boldsymbol{q})\left(\boldsymbol{a} - \dot{\boldsymbol{J}}(\boldsymbol{q})\dot{\boldsymbol{q}}\right) + \boldsymbol{\phi}_n , \tag{3.69}$$

where

$$\boldsymbol{J}^{\dagger} = \boldsymbol{W}^{-1}\boldsymbol{J}^{\mathrm{T}}\left(\boldsymbol{J}\boldsymbol{W}^{-1}\boldsymbol{J}^{\mathrm{T}}\right)^{-1} \tag{3.70}$$

denotes the right pseudo-inverse of $\boldsymbol{J}$ weighted by the positive definite $(n \times n)$ matrix $\boldsymbol{W}$. Also, in (3.69), $\boldsymbol{\phi}_n$ denotes a joint acceleration vector lying in the null space of $\boldsymbol{J}$ which is available for redundancy resolution. In fact, plugging (3.69) in (2.8) gives

$$\ddot{\boldsymbol{q}} = \boldsymbol{J}^{\dagger}(\boldsymbol{q})\left(\boldsymbol{a} - \dot{\boldsymbol{J}}(\boldsymbol{q})\dot{\boldsymbol{q}}\right) + \boldsymbol{\phi}_n . \tag{3.71}$$

Then, premultiplying both sides of (3.71) by $\boldsymbol{J}$, accounting for (2.10), and observing that $\boldsymbol{J}\boldsymbol{J}^{\dagger} = \boldsymbol{I}$, $\boldsymbol{J}\boldsymbol{\phi}_n = \boldsymbol{0}$, yields the same end-effector resolved acceleration as in (2.12); $\boldsymbol{I}$ denotes the $(n \times n)$ identity matrix.

The matrix projecting arbitrary joint accelerations into the null space of $\boldsymbol{J}$ is given by $(\boldsymbol{I} - \boldsymbol{J}^{\dagger}\boldsymbol{J})$, no matter what choice is made for the weighting matrix $\boldsymbol{W}$ in (3.70). Therefore, it is significant to choose $\boldsymbol{W}$ so that the redundancy resolution scheme for motion control should not be altered when interaction with the environment occurs. To this purpose, from (2.6) the joint accelerations induced by the external end-effector force and moment are given by

$$\ddot{\boldsymbol{q}}_e = -\boldsymbol{B}^{-1}(\boldsymbol{q})\boldsymbol{J}^{\mathrm{T}}(\boldsymbol{q})\boldsymbol{h} . \tag{3.72}$$

Projecting these accelerations in the null space of the Jacobian gives

$$\ddot{\boldsymbol{q}}_{en} = -\left(\boldsymbol{I} - \boldsymbol{J}^{\dagger}(\boldsymbol{q})\boldsymbol{J}(\boldsymbol{q})\right)\boldsymbol{B}^{-1}(\boldsymbol{q})\boldsymbol{J}^{\mathrm{T}}(\boldsymbol{q})\boldsymbol{h} . \tag{3.73}$$

Choosing $\boldsymbol{W} = \boldsymbol{B}$ in (3.70) and plugging the resulting $\boldsymbol{J}^{\dagger}$ in (3.73) yields $\ddot{\boldsymbol{q}}_{en} = \boldsymbol{0}$, meaning that the external force and moment produce null space joint accelerations. Therefore, in view of this choice, in the design of the joint resolved acceleration in (3.69) the vector $\boldsymbol{\phi}_n$ can be used to solve redundancy independently of the occurrence of interaction with the environment. The matrix

$$\boldsymbol{J}^{\dagger} = \boldsymbol{B}^{-1}\boldsymbol{J}^{\mathrm{T}}\left(\boldsymbol{J}\boldsymbol{B}^{-1}\boldsymbol{J}^{\mathrm{T}}\right)^{-1} \tag{3.74}$$

weighted by the inertia matrix is termed *dynamically consistent pseudo-inverse* of the Jacobian matrix [36]. With this choice, and only with it, as shown in [36], it is possible to meet the requirement on the dynamic separation between internal motions and end-effector forces.

3.5.1 Stabilization of internal motion

The next step consists of designing a redundancy resolution control in terms of the null space joint accelerations $\boldsymbol{\phi}_n$ in (3.69). To this purpose, $\boldsymbol{\phi}_n$ shall be chosen so as to ensure stabilization of the null space motion and possibly optimization of an additional task function. As proposed in [58], let

$$\boldsymbol{e}_n = \left(\boldsymbol{I} - \boldsymbol{J}^\dagger(\boldsymbol{q})\boldsymbol{J}(\boldsymbol{q})\right)(\boldsymbol{\beta} - \dot{\boldsymbol{q}}) \tag{3.75}$$

denote the null space velocity error where $\boldsymbol{\beta}$ is a joint velocity vector which is available for redundancy resolution. The goal is to make $\boldsymbol{e}_n$ asymptotically converge to zero. Taking the time derivative of (3.75) and using (3.71) gives the null space dynamics, i.e.

$$\begin{aligned}\dot{\boldsymbol{e}}_n &= \left(\boldsymbol{I} - \boldsymbol{J}^\dagger(\boldsymbol{q})\boldsymbol{J}(\boldsymbol{q})\right)(\dot{\boldsymbol{\beta}} - \boldsymbol{\phi}_n) \\ &\quad - \left(\dot{\boldsymbol{J}}^\dagger(\boldsymbol{q})\boldsymbol{J}(\boldsymbol{q}) + \boldsymbol{J}^\dagger(\boldsymbol{q})\dot{\boldsymbol{J}}(\boldsymbol{q})\right)(\boldsymbol{\beta} - \dot{\boldsymbol{q}}) \,,\end{aligned} \tag{3.76}$$

where $\dot{\boldsymbol{J}}^\dagger$ is a shortcut notation for the time derivative of $\boldsymbol{J}^\dagger$.

Consider the Lyapunov function candidate

$$\mathcal{V} = \frac{1}{2}\boldsymbol{e}_n^{\mathrm{T}}\boldsymbol{B}(\boldsymbol{q})\boldsymbol{e}_n \,. \tag{3.77}$$

Computing the time derivative of (3.77) along the trajectories of system (3.77) yields

$$\dot{\mathcal{V}} = \frac{1}{2}\boldsymbol{e}_n^{\mathrm{T}}\dot{\boldsymbol{B}}\boldsymbol{e}_n + \boldsymbol{e}_n^{\mathrm{T}}\boldsymbol{B}\left(\dot{\boldsymbol{\beta}} - \boldsymbol{\phi}_n - \dot{\boldsymbol{J}}^\dagger\boldsymbol{J}(\boldsymbol{\beta} - \dot{\boldsymbol{q}})\right) \,, \tag{3.78}$$

where the dependence on $\boldsymbol{q}$ has been dropped off, and the identity

$$\boldsymbol{e}_n^{\mathrm{T}}\boldsymbol{B}\boldsymbol{J}^\dagger = \boldsymbol{0}^{\mathrm{T}} \tag{3.79}$$

has been exploited.

Choosing

$$\boldsymbol{\phi}_n = (\boldsymbol{I} - \boldsymbol{J}^\dagger\boldsymbol{J})\left(\dot{\boldsymbol{\beta}} - \dot{\boldsymbol{J}}^\dagger\boldsymbol{J}(\boldsymbol{\beta} - \dot{\boldsymbol{q}}) + \boldsymbol{B}^{-1}(\boldsymbol{K}_n\boldsymbol{e}_n + \boldsymbol{C}\boldsymbol{e}_n)\right) \,, \tag{3.80}$$

where $\boldsymbol{K}_n$ is a positive definite matrix, and folding it into (3.78) gives

$$\dot{\mathcal{V}} = \frac{1}{2}\boldsymbol{e}_n^{\mathrm{T}}(\dot{\boldsymbol{B}} - 2\boldsymbol{C})\boldsymbol{e}_n - \boldsymbol{e}_n^{\mathrm{T}}\boldsymbol{K}_n\boldsymbol{e}_n = -\boldsymbol{e}_n^{\mathrm{T}}\boldsymbol{K}_n\boldsymbol{e}_n \leq 0 \tag{3.81}$$

owing to (3.79) and the skew-symmetry of the matrix $\dot{\boldsymbol{B}} - 2\boldsymbol{C}$. Therefore, it can be concluded that the choice (3.80) gives a negative definite $\dot{\mathcal{V}}$, and thus $\boldsymbol{e}_n \to \boldsymbol{0}$ asymptotically.

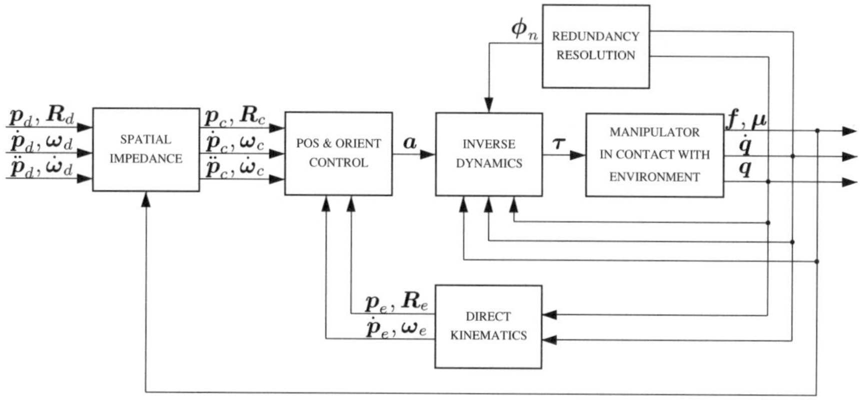

Fig. 3.12. Spatial impedance control with redundancy resolution

Regarding the redundancy resolution, a typical choice for β is [59]

$$\beta = k_\beta B^{-1} \left(\frac{\partial w(q)}{\partial q} \right), \tag{3.82}$$

where k_β is a signed scalar and $w(q)$ is an additional task function that can be locally optimized.

A block diagram summarizing the overall inverse dynamics control with redundancy resolution is sketched in Fig. 3.12. The inverse dynamics control law gives τ as in (3.60) with ϕ as in (3.69). The position and orientation control gives the resolved acceleration a as in (2.14) with a_p as in (3.61) and a_o as any of (3.63),(3.65),(3.66), (3.67). The direct kinematics gives the actual end-effector position and orientation as in (2.1),(2.2a), and the linear and angular velocity as in (2.3),(2.4). Finally, the redundancy resolution scheme gives ϕ_n as in (3.80) with β as in (3.82). Obviously, if a non-redundant manipulator is of concern, then $\phi_n = 0$.

3.5.2 Experiments

In the case study carried out to test the effectiveness of the proposed task space impedance controller with redundancy resolution, the seven-joint manipulator available in the PRISMA Lab is used to interact with an environment constituted by a cardboard box. The translational stiffness at the contact between the end effector and the surface is of the order of 5000 N/m, while the rotational stiffness for small angles is of the order of 15 Nm/rad.

The task in the experiment consists of four phases; namely, reconfiguring the manipulator, approaching the surface, staying in contact, and leaving the surface. To begin, the additional task function in (3.82) has been chosen as

$$w(q) = \frac{1}{2}(q_3 - q_{3d})^2$$

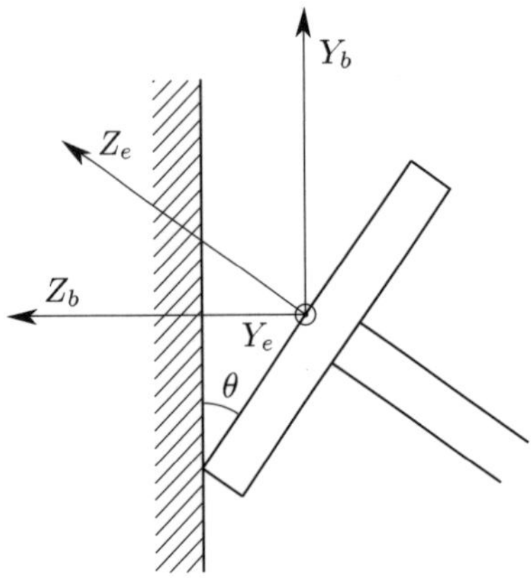

Fig. 3.13. Disk in contact with surface

where q_3 is the elbow joint and q_{3d} is a desired trajectory from the initial value of q_3 to the final value of 1.1 rad in a time of 4 s with a fifth-order interpolating polynomial, with null initial and final velocity and acceleration. This function is aimed at reconfiguring the manipulator in a more dexterous posture before contacting the surface.

After a lapse of 4 s, the disk is taken in contact with the surface at an angle $\theta = 7\pi/36$ rad; see Fig. 3.13 where the orientation of the base and end-effector frames is depicted. The desired end-effector position is required to make a straight-line motion with a horizontal displacement of 0.08 m along the Z_b axis of the base frame. The trajectory along the path is generated according to a fifth-order interpolating polynomial with null initial and final velocities and accelerations, and a duration of 2 s. The desired end-effector orientation is required to remain constant during the task. The surface is placed (vertically) in the X_bY_b-plane of the base frame in such a way as to obstruct the desired end-effector motion, both for the translational part and the rotational part. After a lapse of 13 s in contact, the end-effector motion is commanded back to the initial position with a duration of 4 s.

The parameters of the translational impedance (3.9) have been set to

$$\boldsymbol{M}_p = 16\boldsymbol{I}, \quad \boldsymbol{D}_p = \mathrm{diag}\{800, 800, 250\}, \quad \boldsymbol{K}_p = \mathrm{diag}\{1300, 1300, 800\}\,,$$

while the parameters of the rotational impedance (3.52) have been set to

$$\boldsymbol{M}_o = 0.7\boldsymbol{I}, \quad \boldsymbol{D}_o = 4\boldsymbol{I}, \quad \boldsymbol{K}_o = 2.5\boldsymbol{I}\,.$$

The gains of the inner motion control loop in (3.61),(3.67) have been tuned to the values

$$\boldsymbol{K}_{Pp} = 2250\boldsymbol{I}, \quad \boldsymbol{K}_{Dp} = 70\boldsymbol{I}$$

for the rotational part, and to the values

$$K_{Po} = 4000I, \quad K_{Do} = 75I$$

for the rotational part. The gains of the redundancy resolution control in (3.80),(3.82) have been set to

$$K_n = 20I, \quad k_\beta = 250 .$$

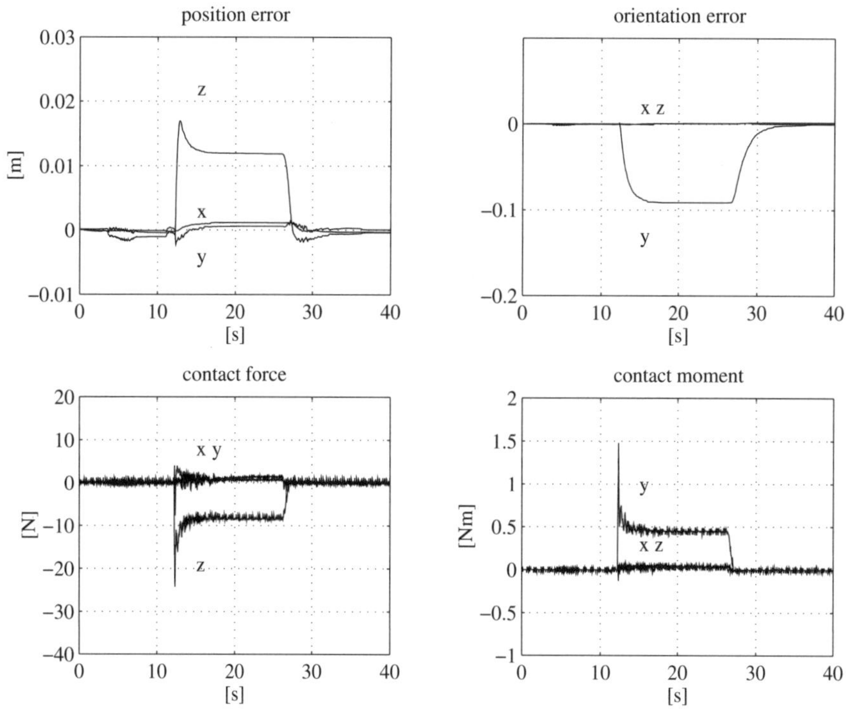

Fig. 3.14. Experimental results under six-DOF impedance control based on quaternion with redundancy resolution

The results in Fig. 3.14 show the effectiveness of the six-DOF impedance control with redundancy resolution. During the reconfiguration (8 s), the components of the position error $\Delta p_{de} = p_d - p_e$ between Σ_d and Σ_e and of the orientation error ${}^e\epsilon_{de}$ between Σ_d and Σ_e are practically zero, meaning that the dynamics of the null space motion does not disturb the end-effector motion. Such error remains small during the approach (2 s). During the contact (13 s), the component of the position error along the Z_b-axis significantly deviates from zero, as expected; as for the orientation error, the component of the orientation error along the Y_e-axis significantly deviates from zero since Σ_e has to rotate about Y_e in order to comply with the surface. Also, in view of the imposed task, a prevailing component of the contact force can

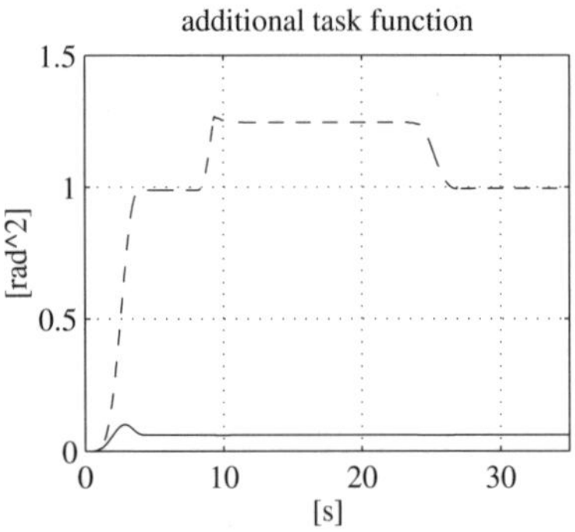

Fig. 3.15. Additional task function in the fifth case study

be observed along the Z_b-axis after the contact, whereas the sole component of the contact moment about the Y_e-axis is significant, as expected. During takeoff (4 s), both the errors and contact force and moment return to zero.

The same task has been executed again for the impedance control without redundancy resolution ($k_\beta = 0$). The performance in terms of the contact between the end effector and the surface is the same as above, since the additional task does not interfere with the primary interaction task; hence, the time history of the relevant quantities is omitted for brevity. Nevertheless, a comparison between the two cases in Fig. 3.15 shows that the task function is successfully optimized when redundancy is exploited (solid) other than when redundancy is not exploited (dashed).

4. Task Space Force Control

In many applications, the task to be executed by the robot manipulator requires the end effector to exert prescribed values of force and moment on the environment, while following an assigned trajectory. Therefore, the problem of devising a control strategy able to safely manage both interaction and motion in the Cartesian space arises. This is the main topic of the chapter, where the hybrid paradigm is reviewed and compared with the parallel approach. The latter is formulated for the six-DOF case, to achieve tracking of position and orientation along unconstrained directions, as well as regulation of both force and moment along constrained directions, during the interaction with a compliant environment.

4.1 Direct Force Control

In the previous chapter, an indirect control of the contact force has been achieved by suitably using force feedback to control the end-effector motion. In this way, it is possible to ensure limited values of the contact force for a rough given estimate of the environment stiffness. Certain interaction tasks, however, do require the fulfillment of a precise value of the contact force. This would be possible, in theory, by tuning the active compliance control action and selecting a proper desired location for the end effector; such a strategy would be effective only on the assumption that accurate estimates of contact stiffness and of environment location are available.

A radically different approach consists in designing a *direct force control* which operates on a force error between the desired and the measured values. On the other hand, in the previous chapter it has been emphasized that even impedance control, which does not aim at achieving a desired force, needs contact force measurements in order to have a dynamically decoupled end-effector behaviour and gain robustness against joint torque disturbances. Therefore, contact force measurements are fully exploited hereafter to design direct force control.

Since the early 80's, direct force control attracted the interest of many researchers, leading to a huge quantity of literature in the field. Throughout the last decade two main approaches have been emerging from amongst the others, namely the *hybrid control* and the *parallel control*. They have substantially different characteristics but also different areas of applicability—as usual, a solution which works for every problem does not exist.

4.1.1 Hybrid approach

Two different formulations of hybrid control can be found in the literature, i.e., the *geometrical formulation* and the *analytical formulation*.

The basic hybrid position/force control scheme classifiable as a geometrical approach was proposed in [61] and it was based on the ideas of *natural constraints* and *artificial constraints* introduced in [51] to model an interaction task from a geometrical point of view. The former are the degrees-of-freedom of the task space along which the environment imposes either a position or a force constraint on the robot's end effector. The latter are the reference values for the degrees-of-freedom of the task space not subject to natural constraints. The approach is geometrical since the directions of the natural constraints are orthogonal to the directions of the artificial constraints. It is clear that only artificial variables can be controlled.

Among the artificial constraints, a distinction between the motion controlled variables and the force controlled variables should to be introduced. The control algorithm was based on the use of binary *selection matrices*, complementary to each other, in order to avoid interaction between the motion control loop and the force control loop.

Later, in [48], it was shown that the concept of task space decomposition cannot be, in general, included in the hybrid control based on selection matrices due to its non-invariant nature with respect to translating the reference frame origin or changing the length unit. The authors solved the problem of non-invariance by substituting the concept of orthogonality with the more general concept of reciprocity, and selection matrices with kinetostatic filtering.

More recent works showed, e.g. [1, 8], that selection matrices can still be used for invariant hybrid position/force control if a suitable frame can be defined, in which position and force controlled directions can be clearly identified.

The analytical hybrid approach to constraint motion formulation has been proposed in many papers, e.g. [72, 52], where the problem of separation between force controlled and motion controlled variables was solved by referring to the equation of the constraint surface leading to a reduction of the dynamic model of the system, by reducing its degrees-of-freedom, thanks to the use of Lagrangian multipliers technique. On the other hand, such a formulation often leads to loss of geometrical and physical meaning of the variables involved.

Very recently, in [42] the authors showed there exist cases in which the analytical and the geometrical approach are equivalent, e.g. in case of point contact or when the constrained rotational axis intersect each other. They also recognize that the analytical approach is more general but at the same time they suggest using the geometrical approach whenever possible because of its intuitive nature.

4.1.2 Parallel approach

Whenever the hybrid, either geometrical or analytical, approach is used, an accurate model of the environment is required; in fact, the controller structure depends directly on the geometrical or analytical environment model.

This is a critical limitation for two main reasons. First, a model of unstructured environments is very difficult to obtain, hence hybrid control can be adopted only for interaction with structured environments. Then, in the case of interaction with stiff environments, uncertainties on the model can lead to very dangerous situations. For example, if the planned trajectory requires a displacement along a direction which is actually constrained, high contact forces can build up, causing robot or manipulated object breakage.

A technique proposed in [25] to overcome this limitation is the so-called *parallel control*. The basic idea is to control both force and motion in all task directions but in a suitable way to give a higher priority to force regulation. In other words, if a planned motion direction is an actual constrained direction, the force control action prevails over the motion control action, preventing the undesirable effects described above for the case of hybrid control. The force and motion control actions can be designed on the basis of a simplified model of the environment while providing some sort of robustness to uncertainty; it can be asserted that "parallel control is based on rules rather than on models".

To sum up, the main difference between the hybrid approach and the parallel approach is that, in the former, the contact geometry directly influences the structure of the controller, while in the latter, it influences only the references in terms of desired motion and desired force; furthermore, in the control system the actual constrained and unconstrained directions are identified during task execution, by properly using force measurement, without any filtering action.

This provides robustness to uncertainty. In fact, even if an unexpected force along a planned unconstrained direction builds up, the force control reacts to regulate the contact force to zero, which is the desired value along all the planned unconstrained directions. On the other hand, if a desired force is specified along a direction actually unconstrained, only a drift of the end effector along that direction occurs, without any possible dangerous situation.

4.1.3 Force control with inner motion loop

In the previous chapter the advantages of an impedance control based on the inner/outer loop technique have already been pointed out. The same convenience can be usefully exploited for position/force control design in the case of interaction with a compliant environment, and only in this case. In fact, if the manipulator interacts with a rigid surface, the end-effector motion is constrained, in the sense that it looses some degrees-of-freedom and the contact force is a reaction force, and thus it is not related to end-effector displacements. This means that the contact force can be controlled only by directly controlling joint driving torques, because changing the end-effector position along constrained directions does not make sense, if the surface is rigid. This causes many problems of implementation on industrial robots due to the limits of commercially available control architecture. In fact, on an industrial manipulator it is usually possible only to specify, or at the most correct, a desired motion trajectory. Moreover, even for a control architecture allowing the user to specify a desired joint torque, the torque actually available at the motor shaft is different from the desired one, due to the presence of torque disturbances, e.g. joint dry friction. The only way to get a good performance with direct torque control is with the adoption of shaft torque sensors or resorting to special and expensive mechanical joint design [5].

If a standard industrial control architecture is still preferred, the most effective way to implement a force control is the inner/outer loop, and in the case of contact with rigid environments, a simple but effective solution is to use a compliant force sensor, as already experimented in, e.g., [56, 43]. For this reason, hereafter an interaction with a compliant environment is assumed. Furthermore, since a geometrical approach has been utilized until this point to describe six-DOF interaction tasks, a contact geometry is assumed, so that constrained and unconstrained directions can be clearly identified. More complicated cases can be treated only via the analytical formulation, as discussed in the hybrid approach.

Therefore, the following equations will be used to model the compliant environment

$$\boldsymbol{f} = \boldsymbol{K}_f(\boldsymbol{p}_e - \boldsymbol{p}_u) \tag{4.1a}$$

$$^c\boldsymbol{\mu} = {}^c\boldsymbol{K}_\mu {}^c\boldsymbol{\omega}_e \mathrm{d}t \,, \tag{4.1b}$$

where $\boldsymbol{K}_f$ and $^c\boldsymbol{K}_\mu$ are the environment symmetric positive semi-definite translational and rotational stiffness, respectively; $\boldsymbol{p}_u$ is the position of the undeformed environment and $^c\boldsymbol{\omega}_e \mathrm{d}t$ denotes an infinitesimal rotational displacement between Σ_e and the frame Σ_u describing the orientation of the undeformed environment. As usual, the frame Σ_c, to which the moment and rotational displacement are referred, is the *compliant frame*, i.e., the frame the end effector should follow during the interaction task. However, in the case of direct force control, the end effector will follow the trajectory of the compliant frame only along the *constrained directions*, namely the directions along which a force or a moment can arise. On the other hand, it will follow the desired trajectory along the *unconstrained directions*. These results will be achieved by defining suitable references for the inner motion loop via the so-called *parallel composition*.

4.2 Task Space Parallel Control

When a six-DOF interaction task is of concern, the interaction control law should preserve the property of *task geometric consistency*. To this purpose, as for the *task space impedance control*, also for the *task space parallel position/force control* the choice of the representation of end-effector orientation plays a key role. In order to save the above cited property a description in the class of angle/axis representations should be adopted. In particular, unit quaternions will be chosen thanks to their geometrical meaning and because they can globally parameterize the group of rotations $SO(3)$ (see Appendix A for technical details), thus avoiding the occurrence of representation singularities and completely replacing rotation matrices.

According to the inner/outer loop technique, an inner motion loop should be designed and the references to be tracked should be suitably computed, by an outer force loop, in order to ensure force and moment regulation along the constrained directions and tracking of the desired trajectory along the unconstrained directions. It is easy to recognize that the *translational constrained directions* are represented by a basis of $\mathcal{R}(\boldsymbol{K}_f)$, while the *rotational constrained directions* are represented by a basis of $\mathcal{R}({}^c\boldsymbol{K}_\mu)$—notice how the task space is decomposed in the base frame, for the translational degrees-of-freedom and in the compliant frame, for the rotational degrees-of-freedom.

In detail, the six-DOF interaction task is specified as follows. The constant desired force is assigned so that $\boldsymbol{f}_d \in \mathcal{R}(\boldsymbol{K}_f)$ and the constant desired moment is assigned so that ${}^c\boldsymbol{\mu}_d \in \mathcal{R}({}^c\boldsymbol{K}_\mu)$. In a dual way, the desired position trajectory is assigned so that $\boldsymbol{p}_d(t) \in \mathcal{N}(\boldsymbol{K}_f)$ —and it is clear that also the linear velocity and acceleration lies in the null space of $\boldsymbol{K}_f$— while the desired orientation trajectory $\mathcal{Q}_{dc}(t), {}^c\boldsymbol{\omega}_{dc}(t), {}^c\dot{\boldsymbol{\omega}}_{dc}(t)$ is assigned so that ${}^c\boldsymbol{\omega}_{dc} \in \mathcal{N}({}^c\boldsymbol{K}_\mu)$. It is worth pointing out that the desired orientation trajectory is specified as a relative orientation between the desired frame Σ_d and the compliant frame Σ_c, in the sense that the quaternion $\mathcal{Q}_{dc}$ is a rotation about an axis which has to be aligned to an unconstrained direction, which has been defined in the compliant frame.

With reference to the robot dynamic model in (2.6) and the motion control law (3.60), the resolved acceleration $\boldsymbol{a} = [\boldsymbol{a}_p^{\mathrm{T}} \quad \boldsymbol{a}_o^{\mathrm{T}}]^{\mathrm{T}}$ is designed as

$$\boldsymbol{a}_p = \ddot{\boldsymbol{p}}_r + \boldsymbol{K}_{Dp}\Delta\dot{\boldsymbol{p}}_{re} + \boldsymbol{K}_{Pp}\Delta\boldsymbol{p}_{re} \tag{4.2}$$

$$\boldsymbol{a}_o = \dot{\boldsymbol{\omega}}_r + \boldsymbol{K}_{Do}\Delta\boldsymbol{\omega}_{re} + \boldsymbol{K}_{Po}\boldsymbol{\epsilon}_{re}\ , \tag{4.3}$$

where the subscript r denotes the reference frame to be tracked and its position $\boldsymbol{p}_r$ and orientation $\mathcal{Q}_r = \{\eta_r, \boldsymbol{\epsilon}_r\}$ are computed through the technique of the *parallel composition*, explained in the following.

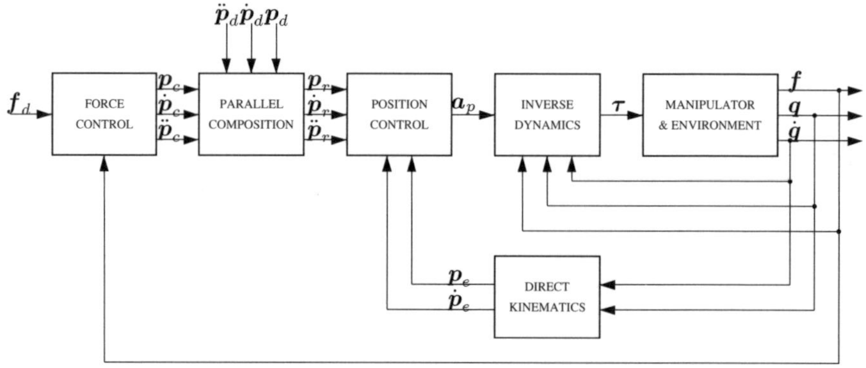

Fig. 4.1. Parallel force and position control

As concerns the translational part, the parallel composition is defined as

$$\boldsymbol{p}_r = \boldsymbol{p}_c + \boldsymbol{p}_d \tag{4.4}$$

$$\dot{\boldsymbol{p}}_r = \dot{\boldsymbol{p}}_c + \dot{\boldsymbol{p}}_d \tag{4.5}$$

$$\ddot{\boldsymbol{p}}_r = \ddot{\boldsymbol{p}}_c + \ddot{\boldsymbol{p}}_d \tag{4.6}$$

being $\boldsymbol{p}_c$ in (4.4) the solution to the differential equation expressing the force control law

$$\boldsymbol{K}_{Af}\ddot{\boldsymbol{p}}_c + \boldsymbol{K}_{Vf}\dot{\boldsymbol{p}}_c = \Delta\boldsymbol{f} \tag{4.7}$$

with $\Delta\boldsymbol{f} = \boldsymbol{f}_d - \boldsymbol{f}$, $\boldsymbol{p}_c(0) = \boldsymbol{0}$, $\boldsymbol{K}_{Af} = k_{Af}\boldsymbol{I}$ and $\boldsymbol{K}_{Vf} = k_{Vf}\boldsymbol{I}$. It is worth pointing out that $\boldsymbol{p}_c$ resulting from integration of (4.7) provides an *integral control* action on the force error $\Delta\boldsymbol{f}$.

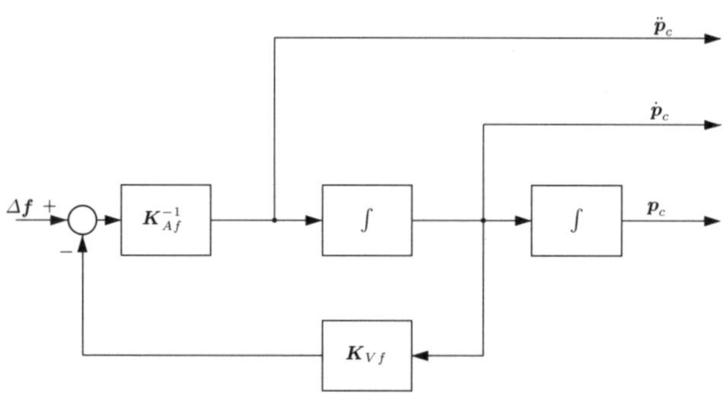

Fig. 4.2. Force controller

The resulting parallel controller for the translational part, originally proposed in [10], is outlined in Fig. 4.1, while in Fig. 4.2 the explicit block scheme of the force controller is depicted, to clearly show the integral action on force error.

As regards the rotational part, the parallel composition is defined as

$$\mathcal{Q}_r = \mathcal{Q}_c * \mathcal{Q}_{dc} \tag{4.8}$$

$$^c\boldsymbol{\omega}_r = {^c\boldsymbol{\omega}_c} + {^c\boldsymbol{\omega}_{dc}} \tag{4.9}$$

$$^c\dot{\boldsymbol{\omega}}_r = {^c\dot{\boldsymbol{\omega}}_c} + {^c\dot{\boldsymbol{\omega}}_{dc}}\ , \tag{4.10}$$

where $\mathcal{Q}_c$, $\boldsymbol{\omega}_c$ and $\dot{\boldsymbol{\omega}}_c$ characterize the rotational motion of the compliant frame Σ_c.

These quantities can be computed once the rotational motion of Σ_c has been computed according to the differential equation expressing the moment controller

$$\boldsymbol{K}_{Am}{^c\dot{\boldsymbol{\omega}}_c} + \boldsymbol{K}_{Vm}{^c\boldsymbol{\omega}_c} = \Delta^c\boldsymbol{\mu}\ , \tag{4.11}$$

with $\Delta^c\boldsymbol{\mu} = {^c\boldsymbol{\mu}_d} - {^c\boldsymbol{\mu}}$, $\mathcal{Q}_c(0) = \mathcal{Q}_u$, $\boldsymbol{K}_{Am} = k_{Am}\boldsymbol{I}$ and $\boldsymbol{K}_{Vm} = k_{Vm}\boldsymbol{I}$. As for the previous force and position control scheme, $\mathcal{Q}_c$ resulting from integration of (4.11) together with the quaternion propagation (A.28a),(A.28b), provides an integral control action on the moment error.

A block diagram of the resulting moment and orientation control scheme, originally proposed in [57], is sketched in Fig. 4.3. This scheme is the counterpart of the force and position control scheme in Fig. 4.1, where the moment control generates the orientation, angular velocity and angular acceleration of Σ_c as in (4.11); these are combined with the desired orientation, angular velocity and angular acceleration as in (4.8),(4.9) and (4.10) to generate the corresponding reference quantities to be input to the orientation control.

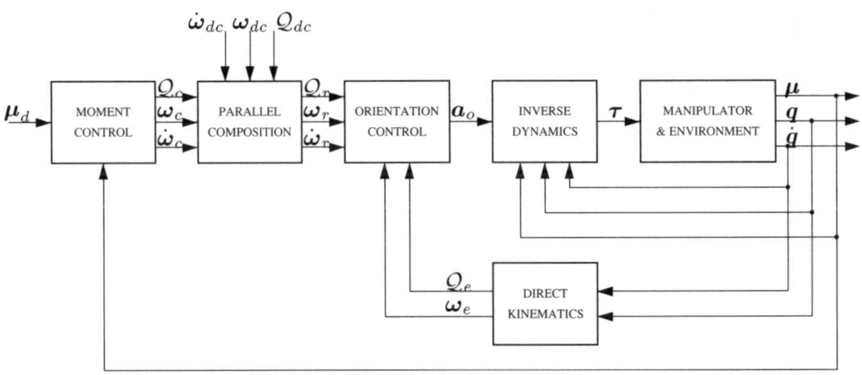

Fig. 4.3. Parallel moment and orientation control

In order to show that the above described interaction task is actually fulfilled, it must be shown the actual force and moment asymptotically tend to the desired values along constrained directions and the actual end-effector translational and rotational trajectories track the desired trajectories along unconstrained directions.

Starting with the translational part, from (4.2),(2.15a) it follows that the actual end-effector trajectory tracks the reference trajectory, and in the hypothesis that the system starts with matched initial conditions, it is $\boldsymbol{p}_e = \boldsymbol{p}_r$ and $\dot{\boldsymbol{p}}_e = \dot{\boldsymbol{p}}_r$. Taking into account the parallel composition rule (4.4),(4.5) leads to

$$\boldsymbol{p}_e = \boldsymbol{p}_d + \boldsymbol{p}_c \tag{4.12}$$

$$\dot{\boldsymbol{p}}_e = \dot{\boldsymbol{p}}_d + \dot{\boldsymbol{p}}_c \,. \tag{4.13}$$

Accounting for the assumption $\boldsymbol{f}_d \in \mathcal{R}(\boldsymbol{K}_f)$ and looking at the equation of the force controller (4.7), implies that $\dot{\boldsymbol{p}}_c, \boldsymbol{p}_c \in \mathcal{N}^{\perp}(\boldsymbol{K}_f)$; hence along the unconstrained directions (basis of $\mathcal{N}(\boldsymbol{K}_f)$) the end-effector trajectory tracks the desired trajectory, because $\dot{\boldsymbol{p}}_c, \boldsymbol{p}_c$ cannot provide any contribution.

Differentiating with respect to time the force controller equation (4.7) and the environment model (4.1a), accounting for (4.13) and considering the assumptions $\dot{\boldsymbol{p}}_d \in \mathcal{N}(\boldsymbol{K}_f)$ and $\boldsymbol{f}_d = \text{const}$, yields

$$\boldsymbol{K}_{Af} \dddot{\boldsymbol{p}}_c + \boldsymbol{K}_{Vf} \ddot{\boldsymbol{p}}_c + \boldsymbol{K}_f \dot{\boldsymbol{p}}_c = \boldsymbol{0} \,. \tag{4.14}$$

This equation implies that $\boldsymbol{K}_f \dot{\boldsymbol{p}}_c \to \boldsymbol{0}$ provided that $\boldsymbol{K}_{Af}$ and $\boldsymbol{K}_{Vf}$ are positive definite matrix gains. Since $\boldsymbol{K}_f$ is positive semi-definite and $\dot{\boldsymbol{p}}_c \in \mathcal{N}^{\perp}(\boldsymbol{K}_f)$, it can be concluded that

$$\ddot{\boldsymbol{p}}_c, \dot{\boldsymbol{p}}_c \to \boldsymbol{0} \tag{4.15}$$

as t goes to infinity, and thus, from the equation of the force controller (4.7),

$$\boldsymbol{f} \to \boldsymbol{f}_d \tag{4.16}$$

as t goes to infinity, i.e., regulation of the contact force to the desired value is achieved.

As regards the rotational part of the parallel controller, a similar analysis can be conducted. From (4.3),(2.15b) it follows that the actual end-effector trajectory tracks the reference trajectory, and in the hypothesis that the system starts with matched initial conditions, it is $\mathcal{Q}_e = \mathcal{Q}_r$ and $\boldsymbol{\omega}_e = \boldsymbol{\omega}_r$. Taking into account the parallel composition rule (4.8),(4.9) leads to

$$\mathcal{Q}_e = \mathcal{Q}_c * \mathcal{Q}_{dc} \tag{4.17}$$

$${}^c\boldsymbol{\omega}_e = {}^c\boldsymbol{\omega}_c + {}^c\boldsymbol{\omega}_{dc} \,. \tag{4.18}$$

Accounting for the assumption ${}^c\boldsymbol{\mu}_d \in \mathcal{R}({}^c\boldsymbol{K}_\mu)$ and taking into account the equation of the moment controller (4.11) implies that ${}^c\boldsymbol{\omega}_c \in \mathcal{N}^{\perp}({}^c\boldsymbol{K}_\mu)$. Hence, along the unconstrained directions (basis of $\mathcal{N}({}^c\boldsymbol{K}_\mu)$) the end-effector trajectory tracks the desired trajectory, because ${}^c\boldsymbol{\omega}_c$ cannot provide any contribution.

Differentiating with respect to time the moment controller equation (4.11) and the environment model (4.1b), accounting for (4.18) and considering the assumptions ${}^c\boldsymbol{\omega}_{dc} \in \mathcal{N}({}^c\boldsymbol{K}_\mu)$ and ${}^c\boldsymbol{\mu}_d = \text{const}$, yields

$$\boldsymbol{K}_{Am} {}^c\ddot{\boldsymbol{\omega}}_c + \boldsymbol{K}_{Vm} {}^c\dot{\boldsymbol{\omega}}_c + {}^c\boldsymbol{K}_\mu {}^c\boldsymbol{\omega}_c = \boldsymbol{0} \,. \tag{4.19}$$

This equation implies that ${}^c\boldsymbol{K}_\mu {}^c\boldsymbol{\omega}_c \to \boldsymbol{0}$ provided that $\boldsymbol{K}_{Am}$ and $\boldsymbol{K}_{Vm}$ are positive definite matrix gains. Since ${}^c\boldsymbol{K}_\mu$ is positive semi-definite and ${}^c\boldsymbol{\omega}_c \in \mathcal{N}^\perp({}^c\boldsymbol{K}_\mu)$, it can be concluded that

$$ {}^c\dot{\boldsymbol{\omega}}_c,\ {}^c\boldsymbol{\omega}_c \to \boldsymbol{0} \tag{4.20} $$

as t goes to infinity, and thus, from the equation of the moment controller (4.11),

$$ {}^c\boldsymbol{\mu} \to {}^c\boldsymbol{\mu}_d \tag{4.21} $$

as t goes to infinity, i.e. regulation of the contact moment to the desired value is achieved.

Notice that the convergence of the control scheme, both for the translational part and for the rotational part, does not require the exact knowledge of the elements of the stiffness matrices, but only of their structure (null and range subspaces). Moreover, if the desired trajectory has a constant non-null component along a constrained direction ($\mathcal{N}^\perp(\boldsymbol{K}_f)$, $\mathcal{N}^\perp({}^c\boldsymbol{K}_\mu)$), the parallel control reacts to it as to disturbance, force or moment, and regulation is still ensured, owing to the integral action. On the other hand, a non-null component along an unconstrained direction ($\mathcal{R}^\perp(\boldsymbol{K}_f)$, $\mathcal{R}^\perp({}^c\boldsymbol{K}_\mu)$) of the desired force, or moment, generates a velocity along that direction. However, this cannot be considered a dangerous situation, since the robot end effector moves in free space along those directions. In this sense the robustness characteristics of the parallel control approach should be seen.

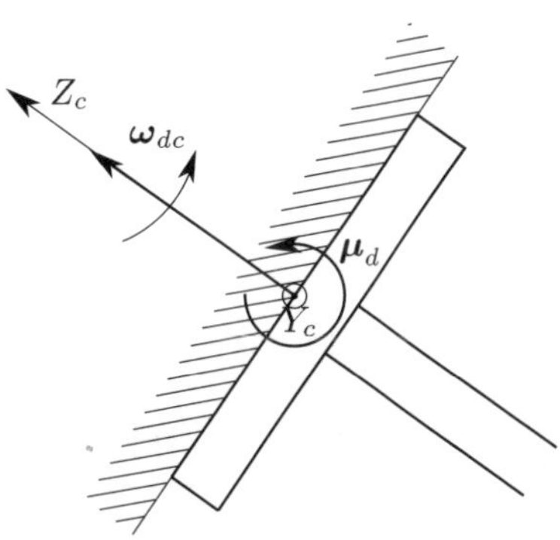

Fig. 4.4. Geometry of contact

4.2.1 Experiments

The above force and motion control scheme has been tested in experiments on the six-joint industrial robot described in Section 1.4 endowed with the force/torque sensor.

The end effector is that reported in Fig. 1.3. The disk is in contact with a plexiglas planar surface such that the approach axis is aligned with the normal to the surface. A schematic description of the contact geometry is depicted in Fig. 4.4.

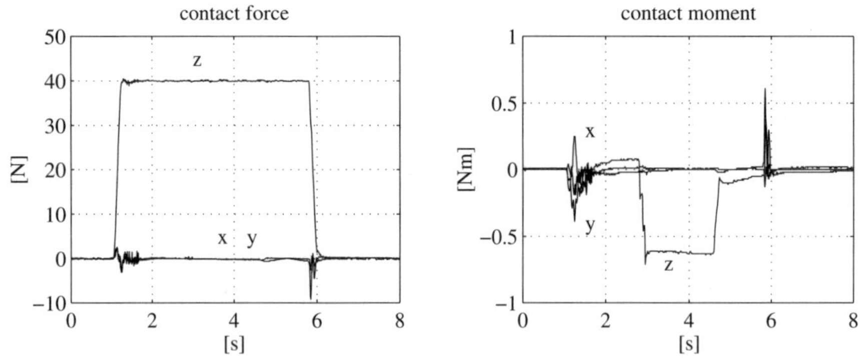

Fig. 4.5. Experimental results under force and moment control

First case study: Force and moment control. The disk is placed in contact with the surface through a rubber coat (to increase the contact friction) so that the rotation about the approach axis is constrained as well as the rotations about the other two axes. In this way, the translational and rotational stiffness are full-rank matrices, i.e.

$$\boldsymbol{K}_f = \mathrm{diag}\{10^3, 10^3, 10^4\}\,\mathrm{N/m}$$

and

$$^c\boldsymbol{K}_\mu = \mathrm{diag}\{20, 20, 40\}\,\mathrm{Nm/rad}\,.$$

The compliant frame is determined so that its Z_c-axis is aligned with the normal to the surface, hence, since the surface is placed horizontally, the Z_c-axis is antialigned to the Z_b-axis. The disk is initially placed in contact and aligned with the surface.

The interaction task consists in keeping a constant position and orientation, and at the same time applying a force normal to the surface and a moment about the approach axis. At the time instant of 1 s a force of 40 N along the Z_b axis is commanded while the desired moment is zero, after a lapse of 1.5 s the desired moment about the approach axis is taken to a value of -0.7 Nm. Both force and moment are then taken back to zero. The variations in the force and moment references are generated according to a 5-th order polynomial with null initial and final first and second time derivatives.

A full six-DOF inverse dynamics control is used, where the linear acceleration is taken as a pure position control with the gains in (4.2) as

$$K_{Dp} = 65I, \quad K_{Pp} = 2500I .$$

On the other hand, the angular acceleration is chosen as in (4.3) and the gains have been set at

$$K_{Do} = 65I, \quad K_{Do} = 4500I .$$

The gains in the force controller have been set at

$$K_{Af} = 8I, \quad K_{Vf} = 650I ,$$

while the gains of the moment control action have been tuned to

$$K_{Am} = I, \quad K_{Vm} = 25I .$$

The results in Fig. 4.5, in terms of the time history of force and contact moment, show that regulation of both force and moment is successfully achieved. The peaks of moment components during the initial transient along X_c and Y_c can be ascribed to the slight initial misalignment between the surface and the disk, which is recovered by the moment controller; in fact the desired moment is zero before rising to the desired non-null value.

Second case study: Moment and orientation control. In this case, no material is placed between the disk and the surface, so that the disk can freely rotate about the approach axis sliding on the surface. Thus, the contact is characterized by a rotational stiffness matrix ${}^cK_\mu = \mathrm{diag}\{30, 30, 0\}$ Nm/rad, i.e., the unconstrained motion is described by any rotation about the approach axis. The compliant frame is the same as the first case study.

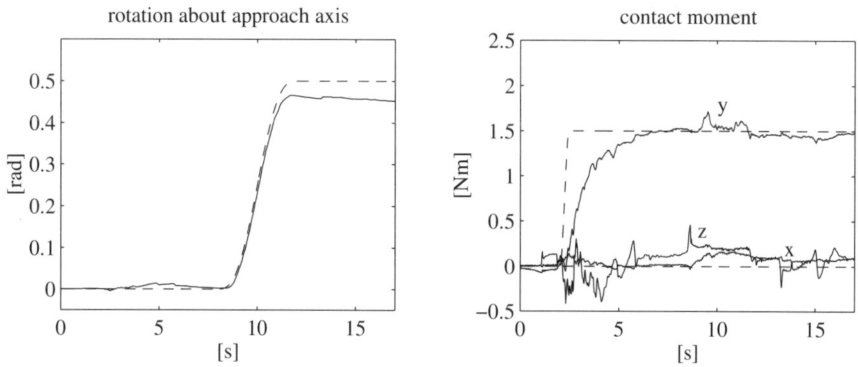

Fig. 4.6. Experimental results under moment and orientation control

The interaction task is as follows. The position is kept constant; at $t = 2\,\mathrm{s}$, the moment is taken to $[0 \quad 1.5 \quad 0]^{\mathrm{T}}$ in Σ_c, according to a 5-th order polynomial with null initial and final first and second time derivatives, and a duration of 0.5 s; the final value is kept constant for the remaining portion of the task. The desired end-effector position is kept constant. After a lapse of 6 s, the desired end-effector orientation is required to make a rotation of 0.5 rad about the approach axis; the trajectory is generated according to a fifth-order interpolating polynomial with null initial and final velocities and accelerations, and a duration of 4 s. The geometry of the contact during the task is depicted in Fig. 4.4 where the desired angular velocity and contact moment are represented.

The same inner motion loop of the previous case study has been implemented, and the force control loop has been opened, while the gains in the moment control action (4.11) have been set at

$$\boldsymbol{K}_{Am} = \boldsymbol{I}, \quad \boldsymbol{K}_{Vm} = 24\boldsymbol{I} .$$

The results are presented in Fig. 4.6 in terms of the time history of the rotation angle about the approach axis of Σ_e and of the three components of the desired (dashed) and the actual (solid) contact moment. It can be recognized that satisfactory tracking of the desired end-effector orientation trajectory is achieved, while the contact moment is successfully regulated to the desired value. Notice that the moment components along X_c and Z_c are affected by contact friction which causes a steady-state deviation on the rotation angle from the desired value; this cannot be recovered since the integral action operates on all the components of the moment error, causing an orientation error also about the unconstrained motion axis.

5. Applications to a Dual-Robot System

This chapter is devoted to presentation of two new approaches for controlling a dual-arm robotic system, namely a loose cooperative strategy and a tight cooperative strategy. The former is based on a modular control architecture which usefully exploits the task interaction control schemes presented in the last two chapters. The latter is aimed at controlling the interaction of an object, commonly held by the two arms, with a compliant environment.

5.1 Cooperative Manipulation

The adoption of multi-arm systems in lieu of a single robot arm is crucial for the execution of many robotic tasks, e.g., parts mating, transportation of heavy or large objects. In fact, this is just the behaviour commonly adopted by humans for the execution of complex manipulation tasks [17].

In a typical robotic workcell, more than a single arm is available, and thus it is of concern to investigate the possibility of performing machining tasks, e.g., assembly and forming tasks, by involving two robots. Therefore, a *cooperative manipulation* strategy should be sought, in order to suitably control each robot in the workcell for successful task execution. Basically, two types of cooperation can be identified depending on the degree of mutual interaction between the robots of the multi-arm system and on the way the cooperation is realized.

If the common manipulation task is executed by controlling the robots in an independent fashion, i.e., the cooperation is realized only at the *task planning level*, the control strategy can be termed *loose cooperative control*. In such a way, the robots move in a coordinate fashion without any interaction or they interact each other but the controller does not explicitly take into account the exchanged forces.

On the other hand, if the manipulation task is executed by controlling the robots in a coordinated fashion, i.e., the cooperation is realized not only at the task planning level, but also at the *control level*, the control strategy can be called *tight cooperative control*. This is the typical case when the robots tightly grasp a commonly held rigid object, thus creating a closed-kinematic chain.

In this chapter, the dual-arm system based on the industrial robots described in the Section 1.4 is used as a testbed to perform different tasks requiring loose and tight robot cooperation.

5.2 Loose Cooperative Control

A loose cooperative manipulation task is performed when the two robotic arms play complementary roles during task execution. In detail, one robot is operated in the standard mode so that the set of motion planning instructions of native programming language can be exploited, although a pure positional control strategy has to be adopted. The other robot is instead operated in the open mode, in such a way as to achieve a programmable compliance via force/torque feedback and thus mitigate the effects of task planning uncertainties. As compliance strategies, the interaction controllers presented in the two previous chapters will be used, namely, the *task space impedance control* and the *task space force control*.

The solution above described exploits the intrinsic modular structure of the controller. In fact, depending on the task requirements, the user has to modify only specific modules, i.e., the task execution program for one robot and the software module selecting the desired compliance strategy for the other.

This approach for managing cooperative robotic tasks will be applied for the execution of two typical workcell tasks involving the interaction between two parts each of them carried by a robot arm.

5.2.1 Modular control structure

The software resources of the setup described in Section 1.4 are organized in the modular multilayer structure of Fig. 5.1, where four layers are illustrated, both for the standard and the open operating modes, representing the various functions implemented in the controller.

The task planning function is implemented only in the standard operating mode via a PDL 2 software module containing the motion instructions needed to execute a specific task. It is worth pointing out that such a function could be realized also in the open operating mode, if desired, by developing a library of C functions dedicated to task planning.

The trajectory generation layer is in charge of computing the motion trajectories corresponding to the planned task. In the standard operating mode, the references for the joint servos are directly computed by the Servo CPU via an inverse kinematics procedure with joint interpolation from the specifications given in the PDL 2 program. In the open operating mode, the position vector $\boldsymbol{p}_r$ and rotation matrix $\boldsymbol{R}_r$ representing the position and orientation of a reference frame are computed from the force and moment

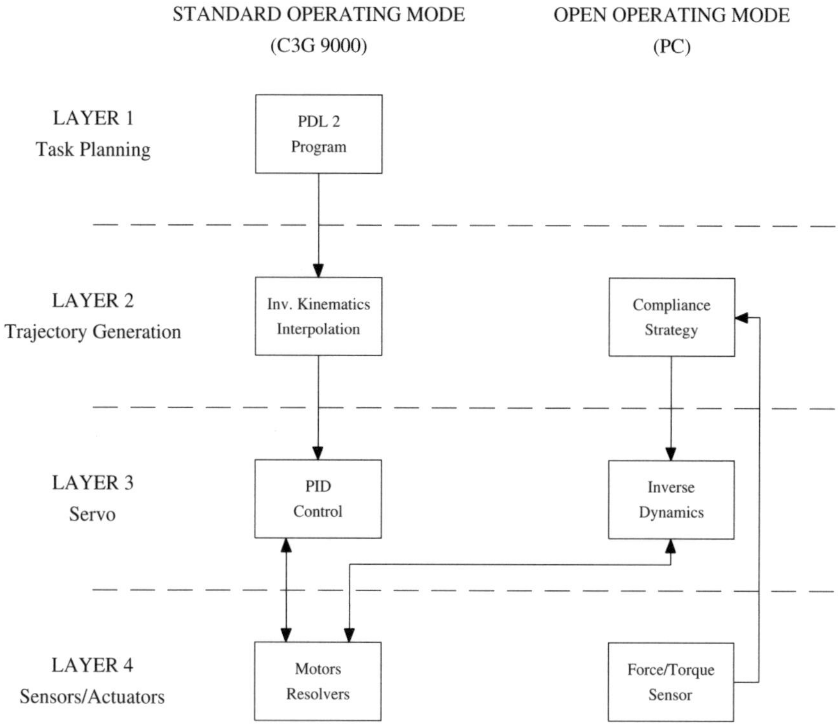

Fig. 5.1. Modular multilayer control structure

measurements on the basis of the compliance strategy specified by the user, and chosen among those described in the two previous chapters of this work.

The servo layer for the standard operating mode implements standard decentralized PID joint position control. In the open operating mode, a task space motion control loop, based on the unit quaternion, is implemented, as described in Chapter 2.

The bottom layer includes joint motors, joint position resolvers and, for the open operating mode, a six-axis force/torque sensor.

In order to program the system for the execution of a given task, the user has to develop the PDL 2 module of the task planning layer and develop the C module implementing the proper compliance strategy in the trajectory generation layer.

5.2.2 Experiments

Two typical machining tasks have been performed to test the efficiency of the modular control structure and the effectiveness of the compliance control strategies of task space impedance and force control, for execution of interaction tasks where the geometry of the contact plays a crucial role. Both compliance control algorithms have been implemented by resorting to the inner/outer loop strategy, hence they are characterized by the same inner loop control law (3.60),(3.61) and (3.67), where the gains have been set at

$$K_{Dp} = 70\boldsymbol{I} \qquad K_{Pp} = 2025\boldsymbol{I}$$

for the position loop, and

$$K_{Do} = 70\boldsymbol{I} \qquad K_{Po} = 5000\boldsymbol{I}$$

for the orientation loop. Notice that a quaternion-based resolved acceleration has been chosen for all the advantages offered by such a representation of orientation, already mentioned many times in this work.

Parts mating task. The archetype of parts mating tasks is the classical peg-in-hole, which can be executed by a dual-arm system, if one robot carries the peg and the other holds the hollow part. It should be clear that the task is successfully executed provided that mating forces are suitably reduced during the insertion to avoid undesirable jamming and wedging. This concept can be brought to fruition by resorting to special mechanical devices such as the Remote Center of Compliance in [70] or the compliant end effectors in [27, 40].

An alternative strategy is to control the robot holding the hollow part to achieve a programmable impedance at the end effector, and the robot carrying the peg to track a trajectory planned to insert the peg into the hole. In this way, the active compliant behaviour imposed on the first robot is in charge of mitigating the effects of imperfect knowledge of the task geometry and unavoidable tracking errors of the second robot. In particular, the seven-joint arm carries the peg while the six-joint arm holds the hollow part.

The task for the seven-joint arm is planned as follows. From a given posture a joint space motion is commanded to reach a suitable intermediate posture which facilitates the subsequent phases of the task; then, a Cartesian space motion along a straight-line path is commanded to drive the tip of the peg in the proximity of the mouth of the hole and align the approach axis of the peg with the axis of the hole, as accurately as possible. Finally, a straight-line motion along the approach axis —typically at a reduced speed with respect to the previous phase— is commanded to realize the insertion. The described task is specified in the following PDL 2 program:

```
PROGRAM insertion CONST spd=3
      dpt=50
VAR   int: JOINTPOS
      pro: POSITION
BEGIN
   --move joints to intermediate posture
   MOVE TO int
   --move tip to proximate pose
   MOVE LINEAR TO pro
```

```
    --set arm speed override at spd %
    $ARM_OVR := spd
    --move tip along approach axis by dpt mm
    MOVE RELATIVE VEC(0,0,dpt) IN TOOL
END insertion
```

where the joint position `int` and the end-effector pose `pro` are taught beforehand, while the constants `spd` and `dpt` are set according to task requirements.

The six-joint arm is controlled using the open operating mode so that the end effector behaves as a mechanical impedance as explained in Chapter 3. To this purpose, the force/torque sensor is mounted at the wrist.

For the task at issue, the position and orientation of the desired frame is taken as a constant, i.e., the six-joint arm is controlled to stay still. Whenever a contact force and/or moment is experienced at the end effector, this reacts according to the programmed impedance where the origin of the desired frame determines the location of the Remote Center of Compliance.

In order to exploit the modularity of the control structure described above, the reference trajectory for the inner loop implemented in the servo layer has to be set equal to the trajectory of the compliant frame, computed in the trajectory generation module by integrating the impedance equations as explained in Chapter 3, i.e., $\boldsymbol{p}_r = \boldsymbol{p}_c$ and $\boldsymbol{R}_r = \boldsymbol{R}_c$.

The peg is a wooden cylinder of 17 mm diameter and 80 mm height, while the hollow part is a wooden block with a hole of 18 mm diameter and 70 mm depth; that is, a 0.5 mm radial tolerance is present during the insertion.

In the first experiment, the six-joint arm is mechanically braked and its end effector is at the position

$$\boldsymbol{p}_0 = [\,885.4 \quad 278.0 \quad 265.6\,]^{\mathrm{T}} \ \mathrm{mm}$$

and orientation

$$\boldsymbol{R}_0 = \begin{bmatrix} \frac{\sqrt{3}}{2} & -\frac{1}{2} & 0 \\ 0 & 0 & 1 \\ -\frac{1}{2} & -\frac{\sqrt{3}}{2} & 0 \end{bmatrix}.$$

An insertion task is programmed in terms of a planned motion for the seven-joint arm described by `spd` = 3 and `dpt` = 50. The proximate position `pro` has been chosen very carefully so as to minimize the misalignment between the peg and the hole axes.

The results are illustrated in Fig. 5.2 in terms of the time history of the three components of mating force and moment. It is easy to recognize that, even though a careful task planning has been accomplished and a complete insertion has not been commanded, large values of force and moment arise, confirming that a pure positional control strategy is always a candidate to fail when performing assembly tasks.

The same task is planned in the second experiment, but this time the six-joint arm is impedance-controlled so that the Remote Center of Compliance is located at

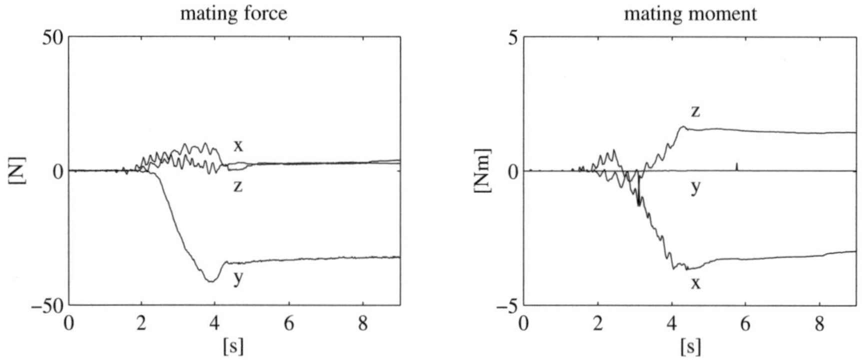

Fig. 5.2. Force and moment in the first experiment of the peg-in-hole task

$$\boldsymbol{p}_d = \boldsymbol{p}_0 - [0 \quad 165 \quad 0]^{\mathrm{T}} \text{ mm}$$

since the insertion direction is along the Y_b-axis of the base frame. Also, it is $\boldsymbol{R}_d = \boldsymbol{R}_0$. The parameters of the impedance equation in (3.9),(3.52) have been respectively set at

$$\begin{aligned} \boldsymbol{M}_p &= \text{diag}\{15, 40, 15\} \\ \boldsymbol{D}_p &= \text{diag}\{300, 950, 300\} \\ \boldsymbol{K}_p &= \text{diag}\{400, 1300, 400\} \end{aligned}$$

for the translational part, and

$$\begin{aligned} \boldsymbol{M}_o &= \text{diag}\{9, 9, 9\} \\ \boldsymbol{D}_o &= \text{diag}\{13.5, 13.5, 13.5\} \\ \boldsymbol{K}_o &= \text{diag}\{1, 1, 1\} \end{aligned}$$

for the rotational part.

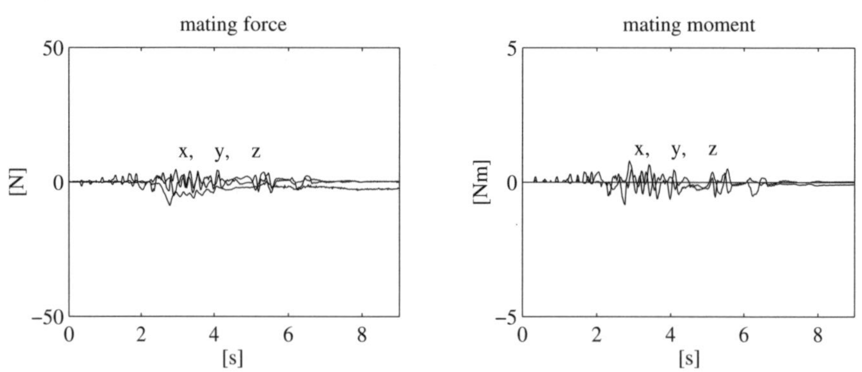

Fig. 5.3. Force and moment in the second experiment of the peg-in-hole task

The results are illustrated in Fig. 5.3 in terms of the time history of the three components of mating force and moment. Remarkably, in this case, the values of force and moment are drastically reduced, indicating that an impedance control strategy is capable to compensate for the errors unavoidably induced by imperfect task planning.

The third and last experiment is aimed at testing the robustness of the proposed strategy in the case of incorrect task planning. An additional misalignment has been intentionally introduced by rotating the end-effector frame of the seven-joint arm by 2 deg about the X_b- and the Z_b-axis of its base frame. Further, the length of the path along the approach axis of the peg has been set to `dpt` = 90, corresponding to a 20 mm overshoot beyond the bottom of the hole. The same impedance parameters as above are chosen.

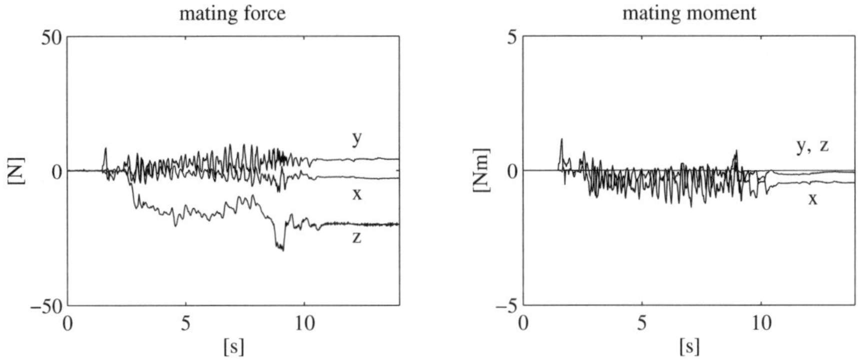

Fig. 5.4. Force and moment in the third experiment of the peg-in-hole task

The results are illustrated in Fig. 5.4 in terms of the time history of the three components of mating force and moment. Interestingly enough, the values of force and moment keep limited despite of the incorrect task planning; notice that they are anyhow smaller than those obtained in the first experiment when an accurate task planning had been attempted, but a position control strategy was adopted. At steady state, nonnull values of force and moment can be observed which are obviously caused by the planned misalignment and overshoot.

Pressure forming task. In the pressure forming task, the seven-joint arm carries a forming tool, while the six-joint arm holds a flat surface to which the tool has to align while exerting a preassigned force and moment.

The task for the seven-joint arm is planned as follows. From a given posture a joint space motion is commanded to reach a suitable intermediate posture which facilitates the subsequent phases of the task; then, a Cartesian space motion along a straight-line path is commanded to drive the disk to the proximity of the surface. Finally, a straight-line motion along the approach axis —typically at a reduced speed compared to the previous phase— is commanded to achieve contact between the disk and the surface. A PDL 2 program similar to that developed for the peg-in-hole task has been employed to execute the described motion.

The six-joint arm is controlled using the open operating mode. In this case the trajectory generation module confers to the workpiece held by the arm a compliance aimed at achieving a preassigned force and moment when in contact with the disk, according to the task space force control law presented in Chapter 4.

In order to realize this compliance strategy, the end-effector position and orientation has to track as accurately as possible the position and orientation of the compliant frame. Hence, the reference trajectory $\boldsymbol{p}_r, \boldsymbol{R}_r$ for the motion control algorithm in the servo layer is computed by solving equations (4.7),(4.11) and then applying the parallel composition.

In the experiment the seven-joint arm carries a disk-shaped tool with a 55 mm diameter, while the six-joint arm holds a wooden surface to which the disk has to align while exerting a desired normal force of 10 N; the alignment of the surface to the disk is achieved by imposing a null desired contact moment.

During the approach of the disk to the surface, the desired force set point along the direction normal to the surface is taken from 0 to 10 N in 8 s according to a 5-th order polynomial; then, after another 20 s the desired force is taken back to zero. This causes a drift motion of the surface along the direction of the desired force (exactly as predicted in Section 4.2), until alignment is achieved (i.e., null contact moment) and the desired force is reached.

The parameters of the force/moment control law (4.7),(4.11) for the six-joint arm have been set at

$$k_{Af} = 20 \qquad k_{Vf} = 400$$

and

$$k_{Am} = 0.81 \qquad k_{Vm} = 30.$$

The results are illustrated in Fig. 5.5 in terms of the time history of the three components of contact force and moment. It can be recognized that after the contact transient, the force along the Z_c-axis normal to the surface reaches the desired value, while all the other components of the force and moment are kept nearly zero. Hence, the surface is aligned with the disk and the desired pressure is maintained. For completeness, in Fig. 5.6 is reported the time history of the rotation executed by the surface to align with the disk, in terms of the scalar and vector part of the quaternion representing the orientation of the compliant frame relative to the initial end-effector frame.

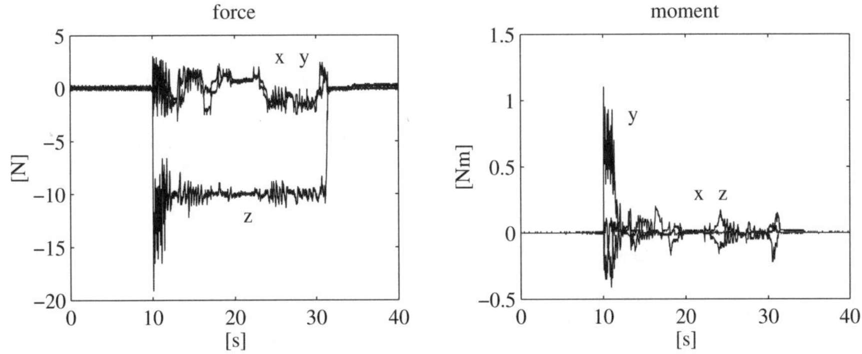

Fig. 5.5. Force and moment in the pressure forming task

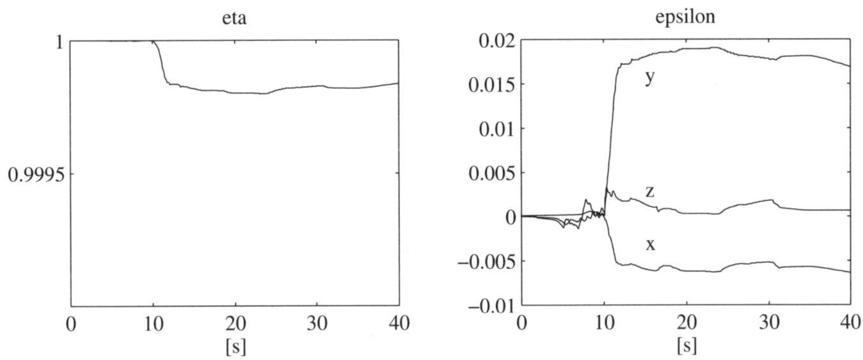

Fig. 5.6. Quaternion representing surface rotation in the pressure forming task

5.3 Tight Cooperative Control

A tight cooperative behaviour is achieved if the motion of the commonly held object is related to the motions of the end effectors of the two robots. This can be done by resorting to the task-oriented formulation for coordinated motion of dual-robot systems developed in [23, 11] and briefly recalled in the next subsection. Moreover, it is worth pointing out that to implement a tight cooperative control strategy, a centralized controller must be used to control both robots. Hence, the open operating mode shall be used to implement the controller and a synchronization technique shall be adopted in order to realize a full coordinated control of the two manipulators.

5.3.1 Task space formulation

Consider a system of two manipulators. For each manipulator ($k = 1, 2$) let Σ_k denote a frame attached to the end effector; its position and orientation are characterized by the (3×1) position vector $\boldsymbol{p}_k$ and the (3×3) rotation matrix $\boldsymbol{R}_k$, respectively. Then, $\mathcal{Q}_k = \{\eta_k, \boldsymbol{\epsilon}_k\}$ represents the unit quaternion corresponding to $\boldsymbol{R}_k$. Let also $\boldsymbol{v}_k = [\,\dot{\boldsymbol{p}}_k^{\mathrm{T}} \quad \boldsymbol{\omega}_k^{\mathrm{T}}\,]^{\mathrm{T}}$ be the (6×1) end-effector (linear and angular) velocity vector. All the quantities are expressed in the common base frame Σ_b.

The location of the commonly held rigid object can be identified by the location of an attached frame Σ_o, called the *object frame*, whose position can be defined as

$$\boldsymbol{p}_o = \frac{1}{2}(\boldsymbol{p}_1 + \boldsymbol{p}_2)\,, \tag{5.1}$$

while the rotation matrix defining its orientation is given by

$$\boldsymbol{R}_o = \boldsymbol{R}_1 {}^1\boldsymbol{R}({}^1\boldsymbol{r}_{21}, \vartheta_{21}/2)\,, \tag{5.2}$$

where ${}^1\boldsymbol{r}_{21}$ and ϑ_{21} are respectively the unit vector and the angle that realize the rotation described by

$${}^1\boldsymbol{R}_2 = \boldsymbol{R}_1^{\mathrm{T}} \boldsymbol{R}_2 \tag{5.3}$$

and ${}^1\boldsymbol{R}({}^1\boldsymbol{r}_{21}, \vartheta_{21}/2)$ is the rotation matrix corresponding to a rotation of $\vartheta_{21}/2$ about the axis ${}^1\boldsymbol{r}_{21}$. Then the absolute orientation can be expressed as

$$\mathcal{Q}_o = \mathcal{Q}_1 * \left\{ \cos\frac{\vartheta_{21}}{4}, \sin\frac{\vartheta_{21}}{4}{}^1\boldsymbol{r}_{21} \right\}\,, \tag{5.4}$$

where $\mathcal{Q}_o$ is the unit quaternion corresponding to $\boldsymbol{R}_o$ and "$*$" denotes the quaternion product; the second factor on the right–hand side of (5.4) is the unit quaternion extracted from ${}^1\boldsymbol{R}({}^1\boldsymbol{r}_{21}, \vartheta_{21}/2)$.

From (5.1),(5.2), the object linear velocity $\dot{\boldsymbol{p}}_o$ and angular velocity $\boldsymbol{\omega}_o$ can be expressed as

$$\boldsymbol{v}_o = \frac{1}{2}(\boldsymbol{v}_1 + \boldsymbol{v}_2)\,, \tag{5.5}$$

where $\boldsymbol{v}_o = [\,\dot{\boldsymbol{p}}_o^{\mathrm{T}} \quad \boldsymbol{\omega}_o^{\mathrm{T}}\,]^{\mathrm{T}}$.

Let $\boldsymbol{f}_k$ and $\boldsymbol{\mu}_k$ $(k = 1, 2)$ respectively denote the (3×1) end-effector force and moment vectors for either manipulator. Then, according to the kinetostatics duality concept [67] applied to (5.5), the object force and moment can be expressed as

$$\boldsymbol{h}_o = \boldsymbol{h}_1 + \boldsymbol{h}_2\,, \tag{5.6}$$

where $\boldsymbol{h}_k = [\,\boldsymbol{f}_k^{\mathrm{T}} \quad \boldsymbol{\mu}_k^{\mathrm{T}}\,]^{\mathrm{T}}$ and $\boldsymbol{h}_o = [\,\boldsymbol{f}_o^{\mathrm{T}} \quad \boldsymbol{\mu}_o^{\mathrm{T}}\,]^{\mathrm{T}}$.

In order to fully describe a coordinated motion, the position and orientation of one manipulator relative to the other is also of concern. The mutual position between the two end effectors is defined as the vector

$$\Delta\boldsymbol{p}_{21} = \boldsymbol{p}_2 - \boldsymbol{p}_1\,. \tag{5.7}$$

The mutual orientation between the two end effectors is defined with reference to Σ_1 in terms of the rotation matrix ${}^1\boldsymbol{R}_2$, and then in terms of the quaternion product

$$\mathcal{Q}_{21} = \mathcal{Q}_1^{-1} * \mathcal{Q}_2 \,, \tag{5.8}$$

where $\mathcal{Q}_1^{-1} = \{\eta_1, -\boldsymbol{\epsilon}_1\}$ is the unit quaternion corresponding to $\boldsymbol{R}_1^{\mathrm{T}}$.

From (5.7),(5.3), the mutual velocity can be expressed as

$$\Delta\boldsymbol{v}_{21} = \boldsymbol{v}_2 - \boldsymbol{v}_1 \,. \tag{5.9}$$

5.3.2 Object-level impedance control

A control strategy for tight cooperative manipulation of an object interacting with the environment can be devised as follows. Two individual motion controllers are developed which guarantee tracking of a reference end-effector position $\boldsymbol{p}_{r,k}$ and orientation $\mathcal{Q}_{r,k}$, as well as of a reference end-effector velocity $\boldsymbol{v}_{r,k}$ $(k = 1, 2)$. Such a reference is generated with a twofold objective; namely, realizing an impedance behaviour at the object level, while assigning a mutual position and orientation between the two end effectors that is compatible with the object geometry.

The first objective can be fulfilled as follows. Let the desired object position $\boldsymbol{p}_d$ and orientation $\mathcal{Q}_d$ (extracted from $\boldsymbol{R}_d$) be assigned with the associated linear and angular velocities and accelerations. Also, the object force and moment can be computed from (5.6) with the end-effector forces and moments available from the wrist force/torque sensors. Then, the translational and rotational impedance equations (3.9),(3.52) are integrated, with input $\boldsymbol{f}_o$ and $\boldsymbol{\mu}_o$, to compute the compliant trajectory, to be followed by the object during the interaction with the environment, $\ddot{\boldsymbol{p}}_c$ and $\dot{\boldsymbol{\omega}}_c$, $\dot{\boldsymbol{p}}_c$ and $\boldsymbol{\omega}_c$, and then $\boldsymbol{p}_c$ and $\mathcal{Q}_c$ via the quaternion propagation (A.28a),(A.28b).

The second objective can be fulfilled by assigning a reference mutual position $\Delta\boldsymbol{p}_{r,21}$ and orientation $\mathcal{Q}_{r,21}$. In particular, $\Delta\boldsymbol{p}_{r,21}$ and $\mathcal{Q}_{r,21}$ are taken as constant and equal to the initial values of $\Delta\boldsymbol{p}_{21}$ in (5.7) and $\mathcal{Q}_{21}$ extracted from (5.3), respectively, that can be computed via the direct kinematics of the two manipulators.

The two objectives are combined by choosing the reference position and orientation for the two end effectors so as to satisfy (5.1),(5.7),(5.4) and (5.8), i.e.,

$$\boldsymbol{p}_{r,1} = \boldsymbol{p}_c - \frac{1}{2}\Delta\boldsymbol{p}_{r,21} \tag{5.10}$$

$$\boldsymbol{p}_{r,2} = \boldsymbol{p}_c + \frac{1}{2}\Delta\boldsymbol{p}_{r,21} \tag{5.11}$$

$$\mathcal{Q}_{r,1} = \mathcal{Q}_c * \left\{\cos\frac{\vartheta_{r,21}}{4}, -\sin\frac{\vartheta_{r,21}}{4}{}^{1}\boldsymbol{r}_{r,21}\right\} \tag{5.12}$$

$$\mathcal{Q}_{r,2} = \mathcal{Q}_{r,1} * \mathcal{Q}_{r,21} \,. \tag{5.13}$$

Further, the reference velocities for the two end effectors are chosen as

$$\boldsymbol{v}_{r,1} = \boldsymbol{v}_c - \frac{1}{2}\Delta\boldsymbol{v}_{r,21} \tag{5.14}$$

$$\boldsymbol{v}_{r,2} = \boldsymbol{v}_c + \frac{1}{2}\Delta\boldsymbol{v}_{r,21} \,, \tag{5.15}$$

where $\boldsymbol{v}_c = [\,\dot{\boldsymbol{p}}_c^{\mathrm{T}} \quad \boldsymbol{\omega}_c^{\mathrm{T}}\,]^{\mathrm{T}}$. Then, the reference accelerations can be computed via an analogous relationship involving accelerations $\dot{\boldsymbol{v}}_c$, $\dot{\boldsymbol{v}}_{r,i}$ and $\Delta\dot{\boldsymbol{v}}_{r,21}$.

The above reference trajectories can be tracked by resorting to an inverse dynamics strategy. The joint driving torques for the two robots can be chosen as ($k = 1, 2$)

$$\boldsymbol{\tau}_k = \boldsymbol{B}_k(\boldsymbol{q}_k)\boldsymbol{J}_k^{-1}(\boldsymbol{q}_k)\big(\boldsymbol{a}_k - \dot{\boldsymbol{J}}_k(\boldsymbol{q}_k, \dot{\boldsymbol{q}}_k)\dot{\boldsymbol{q}}_k\big) + \boldsymbol{n}_k(\boldsymbol{q}_k, \dot{\boldsymbol{q}}_k) + \boldsymbol{J}_k^{\mathrm{T}}(\boldsymbol{q}_k)\boldsymbol{h}_k \,, \tag{5.16}$$

where $\boldsymbol{q}_k$ is the (6×1) vector of joint variables, $\boldsymbol{B}_k$ is the (6×6) symmetric positive definite inertia matrix, $\boldsymbol{n}_k$ is the (6×1) vector of Coriolis, centrifugal and gravity torques, and $\boldsymbol{J}_k$ is the (6×6) (nonsingular) Jacobian matrix. Further in (5.16), $\boldsymbol{a}_k$ is a new control input which can be chosen as $\boldsymbol{a}_k = [\,\boldsymbol{a}_{p,k}^{\mathrm{T}} \quad \boldsymbol{a}_{\epsilon,k}^{\mathrm{T}}\,]^{\mathrm{T}}$ where $\boldsymbol{a}_{p,k}$ and $\boldsymbol{a}_{o,k}$ are designed so as to ensure tracking of $\boldsymbol{p}_{r,k}$ and $\mathcal{Q}_{r,k}$, as well as of $\boldsymbol{v}_{r,k}$, i.e.,

$$\boldsymbol{a}_{p,k} = \ddot{\boldsymbol{p}}_{r,k} + k_{Dp}\Delta\dot{\boldsymbol{p}}_{r,kk} + k_{Pp}\Delta\boldsymbol{p}_{r,kk} \tag{5.17}$$

$$\boldsymbol{a}_{o,k} = \dot{\boldsymbol{\omega}}_{r,k} + k_{Do}\Delta\boldsymbol{\omega}_{r,kk} + k_{Po}\boldsymbol{R}_k{}^{k}\boldsymbol{\epsilon}_{r,kk} \,, \tag{5.18}$$

where $\Delta\boldsymbol{p}_{r,kk} = \boldsymbol{p}_{r,k} - \boldsymbol{p}_k$, ${}^{k}\boldsymbol{\epsilon}_{r,kk}$ is the vector part of $\mathcal{Q}_{r,kk} = \mathcal{Q}_k^{-1} * \mathcal{Q}_{r,k}$, and $\Delta\boldsymbol{\omega}_{r,kk} = \boldsymbol{\omega}_{r,k} - \boldsymbol{\omega}_k$. It is worth remarking that k_{Dp}, k_{Pp} in (5.17) and k_{Do}, k_{Po} in (5.18) are suitable positive feedback gains.

5.3.3 Experiment

In the experiment devoted to testing the proposed tight cooperative control strategy, the two robot end effectors tightly grasp the ends of a wooden bar of 1 m length. At the center of the bar a steel stick is fixed with a wooden disk of 5.5 cm radius at its tip.

The environment is constituted of a cardboard box; the translational stiffness at the contact between the disk and the surface is of the order of 5000 N/m, while the rotational stiffness for small angles is of the order of 15 Nm/rad.

The task in the experiment consists of taking the disk in contact with the surface, which is placed at an unknown distance with an angle of an unknown magnitude. The origin of Σ_o is required to make a desired motion along a straight line with a vertical displacement of -0.275 m along the Z_b-axis of Σ_b. The trajectory along the path is generated according to a 5th-order interpolating polynomial with null initial and final velocities and accelerations, and a duration of 6 s. The desired orientation of the object frame is required to remain constant.

The parameters of the translational part of the impedance equation (3.9) have been set at

$$\boldsymbol{M}_p = \mathrm{diag}\{30, 30, 30\}$$
$$\boldsymbol{D}_p = \mathrm{diag}\{555, 555, 555\}$$
$$\boldsymbol{K}_p = \mathrm{diag}\{1300, 1300, 1300\}$$

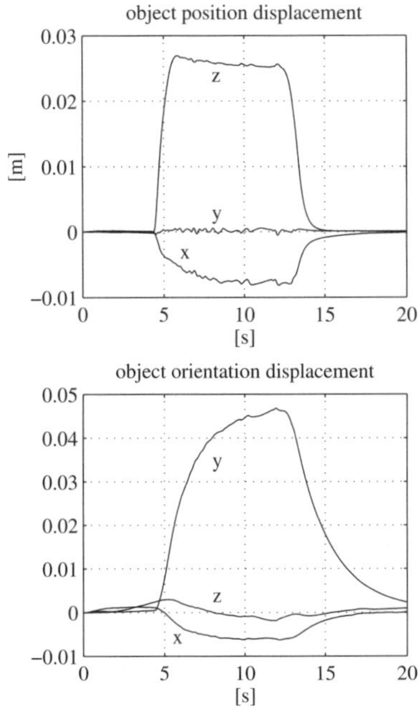

Fig. 5.7. Experiment of tight cooperative control: a–Object position and orientation displacement between Σ_d and Σ_o

while the parameters of the rotational part of the impedance equation (3.52) have been set at

$$\begin{aligned} \boldsymbol{M}_o &= \mathrm{diag}\{10, 2, 10\} \\ \boldsymbol{D}_o &= \mathrm{diag}\{35, 20, 35\} \\ \boldsymbol{K}_o &= \mathrm{diag}\{20, 8, 20\} \end{aligned}$$

Notice that the translational and rotational stiffness matrices have been chosen to ensure a compliant behaviour (limited values of contact force and moment) during the contact, while the damping matrices have been chosen to guarantee a well-damped behaviour.

The feedback gains in (5.17),(5.18) have been set at

$$k_{Dp} = 65, \quad k_{Pp} = 1800$$

for the position loop, and

$$k_{Do} = 65, \quad k_{Po} = 3600$$

for the orientation loop, respectively. Notice that these values differ from those used for the experiments of loose cooperation, because the sampling time had to be increased to 2 ms in order to synchronize the two robot control units.

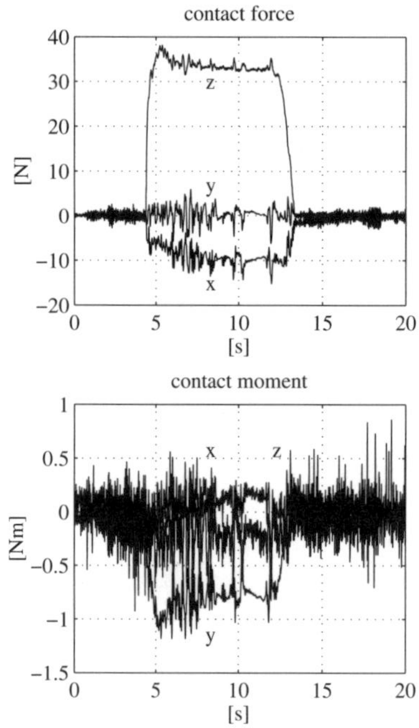

Fig. 5.8. Experiment of tight cooperative control: Contact force and moment acting on the object

From Fig. 5.7, after the contact, the component along the Z_b-axis of the position displacement between the desired frame Σ_d and the object frame Σ_o, expressed in Σ_b, significantly deviates from zero, as expected; a smaller displacement can also be seen for the component along the X_b-axis, due to contact friction. As for the orientation displacement between Σ_o and Σ_d, expressed in Σ_d, only the component along the Y_d-axis significantly deviates from zero since the object frame has to rotate about the Y_d-axis of Σ_d in order to comply with the surface after the contact.

From Fig. 5.8, in view of the imposed task, a prevailing component of the contact force can be observed along the Z_b-axis after the contact, while a significant component along the X_b-axis arises, corresponding to the above position displacement. As for the contact moment, the only nonnegligible component is that along the Y_d-axis of Σ_d, which corresponds to the above orientation displacement. It can be recognized that all the above quantities reach constant steady-state values after the desired motion is stopped. The oscillations on the force and moment can be ascribed to the effects of the commonly held object on the measurements.

6. Conclusion and future research directions

In this work, the topic of control of the interaction between robot manipulators and the external environment for tasks involving six degrees-of-freedom, has been addressed.

The main motivation for this choice has been the evident lack of adequate thoroughness in the related scientific literature about the problems concerning the representation of end-effector orientation to describe the rotational degrees-of-freedom. Most of the authors, in order to extend their results to six dimensional cases, resorted to minimal representations neglecting the problem of representation singularities and more crucially the problem of relating the task geometry to the control system design.

The first step towards the geometric approach has been done in this work, through the analysis of several algorithms for motion control. This comparison has lead to identify the drawbacks of the motion control laws based on the analytical approach via minimal representations. The limits caused by the use of Euler angles have been partially overcome by proposing a suitable alternative analytical formulation of the control algorithm in the operational space. It has been shown that it is possible to completely overcome such drawbacks only by resorting to non-minimal representations of orientation error, via, e.g., the unit quaternion. The properties of this mathematical entity have been deeply analysed in order to understand whether this tool could be used to the purpose of task specification in a way consistent with the task geometry. All the work about motion control, done both from a theoretical and experimental point of view, has provided a great insight into the topic, helpful for the next step towards the design of interaction controllers in the task space.

The two basic strategies for controlling the interaction of a robot manipulator with the environment have been object of study, namely, impedance control and force control.

A class of six-DOF impedance controllers has been proposed. By considering an angle/axis representation of the end-effector orientation displacement, an energy-based argument has lead to deriving a rotational impedance equation at the end effector, which exhibits a physically meaningful behaviour and geometric consistency properties. Among the different angle/axis representations, the unit quaternion has been chosen which avoids the occurrence of representation singularities. The superior performance of the proposed angle/axis-based impedance controllers over two different impedance controllers based on the operational space formulation has been shown both in theory and in practice. An inverse dynamics strategy with contact force and moment measurements has been used in the design of the various controllers which have been made robust thanks to the adoption of an inner loop acting on the end-effector position and orientation error.

Concerning direct force control, motivations for the choice of parallel force/position control, among the various approaches known in the literature for force control problem, have been provided. The parallel formulation has been extended to the case of six-DOF interaction tasks owing to the adoption of unit quaternion, which has allowed the controller to be consistent with the task geometry, without any problem of representation singularities. As in the case of impedance control, the proposed control law has been tested in a number of experiments in order to validate the theoretical findings. The successful implementation on the available industrial setup has been made possible thanks to the adoption of the inner/outer loop strategy.

In order to show the feasibility of the proposed interaction control techniques, the execution of a number of complex robotic tasks has been carried out on the dual-arm system available in the PRISMA Lab. First, two tasks of loose cooperation for the two robots have been conceived, i.e., a parts mating task of peg-in-hole type, and a pressure forming task. The cooperation has been solved at a task planning level, by resorting to a suitable compliance strategy in order to counteract the effects of task planning errors as well as of inaccuracies in the knowledge of the parts dimensions. The satisfactory results have demonstrated the applicability of the proposed control strategy at industrial level, also thanks to the characteristics of modularity inherent in the inner/outer loop technique.

The problem of tight cooperation has also been addressed. The result presented in the last chapter, namely, an impedance controller based on unit quaternion to handle the interaction with an elastic environment of a rigid object commonly held by two manipulators, should be considered only preliminary. In fact, the proposed control law does not explicitly take into account the control of internal forces acting on the object, which may arise during the task execution, caused by tracking errors of inner position loops. The closure of an internal force control loop should be the first step of future research work in this direction, e.g., by adopting an impedance control strategy also for indirect control of internal forces. Some results, have very recently been obtained in this respect [20].

It should be pointed out that, even if all the presented results can be considered very satisfactory, this dissertation does not pretend to be exhaustive about the topic of robot force control, because many other problems are still to be solved and the characteristics of robustness of the proposed control laws can be enhanced. In fact, some advanced techniques, e.g, adaptation with respect to the dynamical parameters and/or environment stiffness, exploitation of passivity characteristics of the robotic system, could be adopted to increase the performance of interaction control algorithms. Some advances of the scientific community in this direction have been recently collected and presented in the monograph [65], but only for 3-DOF interaction tasks. Moreover, interaction tasks with a complex contact geometry may exist, which cannot be addressed by resorting to the proposed task space formulation, hence further work shall be done.

About still more complex tasks, the adoption of a force sensor is certainly not enough to provide the robot with a significant degree of autonomy. In fact, the integration in the robotic system of different heteroceptive sensors is today a big challenge for researchers. Some advances have recently been made in the integration of force and vision feedback for complex assembly tasks, e.g. [43],[7]. Even if these kinds of results have been obtained in an industrial context, they could also be used in the fields of advanced robotics, service robotics, and biorobotics.

References

1. Abbati-Marescotti A., Bonivento C., Melchiorri C. (1990) On the Invariance of the Hybrid Position/Force Control. Journal of Intelligent and Robotic Systems 3:233–250
2. Abrash M. (1994) Zen of Code Optimization: The Ultimate Guide to Writing Software That Pushes PCs to the Limit. Coriolis Group Books, Scottsdale
3. Angeles J. (1988) Rational Kinematics. Springer-Verlag, Ann Arbor
4. Antonelli G., Caccavale F., Chiacchio P. (1999) A Systematic Procedure for the Identification of Dynamic Parameters of Robot Manipulators. Robotica 17:427–435
5. Armstrong-Hèlouvry B., Dupont P., Canudas de Wit C. (1994) A Survay of Models, Analysis Tools and Compensation Methods for the Control of Machines with Friction. Automatica 30:1083–1138
6. Assurance Technologies Inc. (1993) Installation and Operations Manual for F/T. Garner, NC
7. Baeten J., De Schutter J., (1999) Imporving Force Controlled Planar Contour Following Using On-line Eye-in-Hand Vision Based Feedforward. 1999 IEEE/ASME International Conference on Advanced and Intelligent Mechatronics, Atlanta, GA, 902–907
8. Bruyninckx H., De Schutter J. (1996) Specification of Force-Controlled Actions in the "Task Frame Formalism". IEEE Trans. on Robotics and Automation 12:581–589
9. Bruni F., Caccavale F., Natale C., Villani L. (1996) Experiments of Impedance Control on an Industrial Robot Manipulator with Joint Friction. 1996 IEEE Conference on Control Applications, Dearborn, MI, 205–210
10. Bruni F., Natale C. (1995) Sperimentazione di Algoritmi di Controllo dell'Interazione per un Robot Industriale. 39° Convegno Annuale ANIPLA Automazione '95, Bari, I, 439–442
11. Caccavale F., Chiacchio P., Chiaverini S. (2000) Task-Space Regulation of Cooperative Manipulators. Automatica 36:879887
12. Caccavale F., Siciliano B., Villani L. (1999) Robot Impedance Control with Nondiagonal Stiffness. IEEE Transactions on Automatic Control 44:1943–1946
13. Caccavale F., Natale C., Siciliano B., Villani L. (1998) Experiments of Spatial Impedance Control. In: Casals A., de Almeida A.T. (Eds.) Experimental Robotics V, 5th International Symposium on Experimental Robotics, Barcelona, E, June 1997. Springer-Verlag, London, 93–104
14. Caccavale F., Natale C., Siciliano B., Villani L. (1998) Resolved Acceleration Control of Robot Manipulators: A Critical Review with Experiments. Robotica 16:565–573
15. Caccavale F., Natale C., Siciliano B., Villani L. (1998) Control of Two Industrial Robots for Parts Mating. 1998 IEEE International Conference on Control Applications, Trieste, I, Sept. 1998, 562–566

16. Caccavale F., Natale C., Siciliano B., Villani L. (1999) Six-DOF Impedance Control Based on Angle/Axis Representations. IEEE Transactions on Robotics and Automation 15:289–300
17. Caccavale F., Natale C., Siciliano B., Villani L. (2001) Achieving a Cooperative Behaviour in a Dual-Arm Robot System via a Modular Control Structure. Journal of Robotic Systems 18:691–700
18. Caccavale F., Natale C., Siciliano B., Villani L. (2000) Quaternion-Based Impedance Control for Dual-Robot Cooperation. In: Hollerbach J.M., Koditschek D.E. (Eds.) Robotics Research: The Ninth International Symposium, 9th International Symposium of Robotics Research, Snowbird, UT, Oct. 1999. Springer-Verlag, London, 42–52
19. Caccavale F., Natale C., Siciliano B., Villani L. (2000) Geometrically Consistent Impedance Control for Dual-Robot Manipulation. 2000 IEEE International Conference on Robotics and Automation, San Francisco, CA, April 2000, 3874–3879
20. Caccavale F., L. Villani. (2000) Impedance Control for Multi-Arm Manipulation. 39th IEEE Conference on Decision and Control, Sydney, AUS, Dec. 2000, 465–3470
21. Caccavale F., P. Chiacchio. (2000) An experimental set-up for cooperative manipulation based on industrial manipulators. Industrial Robot 27:120–130
22. Canudas de Wit C., Olsson H., Åström K.J., Lischinsky P. (1995) A New Model for Control of Systems with Friction. IEEE Transactions on Robotics and Automation 40:419–425
23. Chiacchio P., Chiaverini S., Siciliano B. (1996) Direct and Inverse Kinematics for Coordinated Motion Tasks of a Two-Manipulator System. ASME Journal of Dynamic Systems, Measurement, and Control 118:691–697
24. Chiaverini S. (1997) Singularity-Robust Task-Priority Redundancy Resolution for Real-Time Kinematic Control of Robot Manipulators. IEEE Transactions on Robotics and Automation 13:398–410
25. Chiaverini S., Sciavicco L. (1993) The Parallel Approach to Force/Position Control of Robotic Manipulators. IEEE Transactions on Robotics and Automation 9:361–373
26. Chou J.C.K. (1992) Quaternion Kinematic and Dynamic Differential Equations. IEEE Transactions on Robotics and Automation 8:53–64
27. Cutkosky M.R., Wright P.K. (1986) Active Control of a Compliant Wrist in Manufacturing Tasks. ASME Journal of Dynamic Systems, Measurement, and Control 108:36–43
28. De Luca A., Oriolo G., Siciliano B. (1992) Robot Redundancy Resolution at the Acceleration Level. Laboratory Robotics and Automation 4:97–106
29. De Schutter J., Bruyninckx H., Zhu W.-H., Spong M.W. (1998) Force Control: A Bird's Eye View. In: B. Siciliano, and K.P. Valavanis (Eds.) Control Problems in Robotics and Automation. Springer-Verlag, London, 1–17
30. De Schutter J., Van Brussel H. (1988) Compliant Robot Motion II. A Control Approach Based on External Control Loops. International Journal of Robotics Research 7:18–33
31. Dogliani F., Magnani G., Sciavicco L. (1993) An Open Architecture Industrial Controller. Newsletter of IEEE Robotics and Automation Society 7:19–21
32. Dombre E., Khalil W. (1988) Modélisation et Commande des Robots. Hermès, Paris
33. Egeland O., Godhavn J.-M. (1994) Passivity-based Adaptive Attitude Control of a Rigid Spacecraft. IEEE Transactions on Automatic Control 39:842–846
34. Fasse E.D., Broenink J.F. (1997) A Spatial Impedance Controller for Robotic Manipulation. IEEE Transactions on Robotics and Automation, 13:546–556

35. Fasse E.D. (1997) On the Spatial Compliance of Robotic Manipulators. ASME Journal of Dynamic Systems, Measurement, and Control 119:839–844
36. Featherstone R., Khatib O. (1997) Load Independence of the Dynamically Consistent Inverse of the Jacobian Matrix. International Journal of Robotics Research 16:168–170
37. Gorinevski D.M., Formalsky A.M., Schneider A.Yu. (1997) Force Control of Robotics Systems. CRC Press, Boca Raton
38. Hennessy J.L., Patterson D.A. (1995) Computer Architecture: A Quantitative Approach. 2nd edn. Morgan Kaufmann Publishers Inc., San Mateo
39. Hogan N. (1985) Impedance control: An Approach to Manipulation: Parts I—III. ASME Journal of Dynamic Systems, Measurement, and Control 107:1–24
40. Hollis R.L., Salcudean S.E., Allan A.P. (1991) A Six-Degree-of-Freedom Magnetically Levitated Variable Compliance Fine-Motion Wrist. IEEE Transactions on Robotics and Automation 7:320–332
41. Hsu P., Hauser J., Sastry S. (1989) Dynamic Control of Redundant Manipulators. Journal of Robotic Systems 6:133–148
42. Jankowski K.P., ElMaraghy H.A. (1996) Constraint Formulation for Invariant Hybrid Position/Force Control of Robots. ASME Journal of Dynamic Systems, Measurement, and Control 118:290–299
43. Joerg S., Langwald J., Natale C., Stelter J., Hirzinger G. (2000) Flexible Automation Using a Multi-Sensor Approach. 2000 IEEE International Conference on Robotics and Automation , San Francisco, CA, April 2000, 3687–3694
44. Khalil H.K. (1996) Nonlinear Systems, 2nd edn. Prentice Hall, Upper Saddle River
45. Khatib O. (1987) A Unified Approach for Motion and Force Control of Robot Manipulators: The Operational Space Formulation. IEEE Journal of Robotics and Automation 3:43–53
46. Lin S.K. (1988) Euler Parameters in Robot Cartesian Control. 1988 IEEE International Conference on Robotics and Automation, Philadelphia, PA, April 1988, pp. 1676–1681
47. Lin S.K. (1989) Singularity of a Nonlinear Feedback Control Scheme for Robots. IEEE Transactions on Systems, Man, and Cybernetics 19:134–139
48. Lipkin H., Duffy J. (1988) Hybrid Twist and Wrench Control for a Robotic Manipulator. ASME Journal of Mechanisms, Transmissions, and Automation in Design 110:138–144
49. Lončarić J. (1987) Normal Forms of Stiffness and Compliance Matrices. IEEE Journal of Robotics and Automation 3:567–572
50. Luh J.Y.S., Walker M.W., Paul R.P.C. (1980) Resolved-Acceleration Control of Mechanical Manipulators. IEEE Transactions on Automatic Control 25:468–474
51. Mason M.T. (1981) Compliance and Force Control for Computer Controlled Manipulators. IEEE Transactions on Systems, Man, and Cybernatics 6:418–432
52. McClamroch N.H., Wang D. (1988) Feedback Stabilization and Tracking of Constrained Robots. IEEE Transactions on Automatic Control 33:419–426
53. Mills J.K., Goldenberg A.A. (1989) Force and Position Control of Manipulators During Constrained Motion Tasks. IEEE Transactions on Robotics and Automation 5:30–46
54. Murray R.M., Li Z., Sastry S.S. (1994) A Mathematical Introduction to Robotic Manipulation. CRC Press, Boca Raton
55. Natale C. (1997) Quaternion-Based Representation of Rigid Bodies' Orientation. PRISMA Technical Report 97-05, Naples

56. Natale C., Koeppe R., Hirzinger G. (2000) A Systematic Design Procedure of Force Controller for Industrial Robots. IEEE/ASME Transactions on Mechatronics 5:122–133
57. Natale C., Siciliano B., Villani L. (1998) Control of Moment and Orientation for a Robot Manipulator in Contact with a Compliant Environment. 1998 IEEE International Conference on Robotics and Automation, Leuven, B, May 1998, 1755–1760
58. Natale C., Siciliano B., Villani L. (1999) Spatial Impedance Control of Redundant Manipulators. 1999 IEEE International Conference on Robotics and Automation, Detroit, MI, May 1999, 1788–1793
59. Nemec B. (1997) Force Control of Redundant Robots 5th IFAC Symposium on Robot Control, Nantes, F, Sept. 1997, 215–220
60. Nijmeijer H., van der Schaft A.J. (1990) Nonlinear Dynamical Control Systems. Springer-Verlag, New York
61. Raibert M.H., Craig J.J. (1981) Hybrid Position/Force Control of Manipulators. ASME Journal of Dynamic Systems, Measurement, and Control 103:126–133
62. Roberson R.E., Schwertassek R. (1988) Dynamics of Multibody Systems. Springer-Verlag, Berlin
63. Sciavicco L., Siciliano B. (2000) Modelling and Control of Robot Manipulators, 2nd edn. Springer, London
64. Shepperd S.W. (1978) Quaternion From Rotation Matrix. AIAA Journal of Guidance and Control 1:223–224
65. Siciliano B., Villani L. (2000) Robot Force Control. Kluwer Academic Publishers, Boston
66. Tecnospazio (1993) Controllore Comau C3G 9000 Aperto – Manuale d'uso, Tecnospazio, Milano
67. Uchiyama M., Dauchez P. (1993) Symmetric Kinematic Formulation and Non-Master/Slave Coordinated Control of Two-Arm Robots. Advanced Robotics 7:361–383
68. Vukobratović M., Nakamura Y. (1993) Force and Contact Control in Robotic Systems. Tutorial at 1993 IEEE International Conference on Robotics and Automation, Atlanta, GA, May 1993
69. Wen J.T.-Y., Kreutz-Delgado K. (1991) The Attitude Control Problem. IEEE Transactions on Automatic Control 36:1148–1162
70. Whitney D.E. (1982) Quasi-static Assembly of Compliantly Supported Rigid Parts. ASME Journal of Dynamic Systems, Measurement, and Control 104:65–77
71. Whitney D.E. (1987) Historical Perspective and State of the Art in Robot Force Control (1987) International Journal of Robotics Research 6:3–14
72. Yoshikawa T. (1987) Dynamic Hybrid Position/Force Control of Robot Manipulators —Description of Hand Constraints and Calculation of Joint Driving Force. IEEE Journal of Robotics and Automation 3:386–392
73. Yuan J.S.-C. (1988) Closed-Loop Manipulator Control Using Quaternion Feedback. IEEE Journal of Robotics and Automation 4:434–440

A. Orientation of a Rigid Body

This appendix is aimed at providing the necessary background on the representation of rigid body orientation in the Cartesian space. Minimal and non-minimal representations are discussed and compared, and the principal properties of the unit quaternion are summarized. Also, the problem of orientation displacement is addressed as a key issue in the feedback control framework.

A.1 Non-minimal Representations

Call Σ_0 a reference frame (the base frame) and Σ_1 a frame attached to the rigid body (the body-fixed frame). *Motions* of the body are described by parameters characterizing the relative position and orientation of the two frames as a function of time. The motion is called a *rotation* when the relative position of the two origins remains fixed. In the 18th century Euler showed that every rotation is equivalent to a rotation about an axis (two parameters) and an angle of rotation about the axis (one parameter), i.e., a rotation is a three-DOF motion—hence the well-known representation of rigid body orientation via the set of the three *Euler angles* [63].

Another relevant property of rotations is that the execution of two successive rotations is not commutative, hence a general rotation cannot be represented by a vector. However, it can be represented by a second order tensor [62]. As a consequence the set of three Euler angles is not a vector.

To this characteristic can be imputed the fact that every three-parameter parametrization of the set of rotations suffers from a singularity, called a *representation singularity*; in fact one may lose smooth dependence of the three parameters as a function of the represented rotation. Therefore, it is worth referring to non-minimal representations, such as rotation matrices and quaternions.

A.1.1 Rotation matrix

Let $\{\boldsymbol{x}, \boldsymbol{y}, \boldsymbol{z}\}$ be the three direction cosines of the base frame and let $\{\boldsymbol{x}', \boldsymbol{y}', \boldsymbol{z}'\}$ be the three direction cosines of the body-fixed frame. The rigid body orientation can be represented with the so called *rotation matrix* or *direction cosine matrix*

$$ {}^0\boldsymbol{R}_1 = [\,\boldsymbol{x}' \quad \boldsymbol{y}' \quad \boldsymbol{z}'\,] \,. \tag{A.1}$$

It can be shown [54] that, if the columns of a rotation matrix $\boldsymbol{R}$ form a right-handed cartesian frame, the set

$$ SO(3) = \left\{ \boldsymbol{R} \in \mathrm{I\!R}^{3\times 3} : \boldsymbol{R}\boldsymbol{R}^{\mathrm{T}} = \boldsymbol{I}, \det\boldsymbol{R} = +1 \right\} \tag{A.2}$$

is a group under the operation of matrix multiplication, called the *special orthogonal*. As expected the set is not a vector space because of the non-commutativity of the matrix product.

The first interpretation of a rotation matrix is as describing the way a single vector transforms when represented with respect to two different vector bases; if ${}^1\boldsymbol{v}$ are the coordinates of a vector $\boldsymbol{v}$ in the body-fixed frame, the coordinates ${}^0\boldsymbol{v}$ in the base frame are computed as

$$ {}^0\boldsymbol{v} = {}^0\boldsymbol{R}_1 {}^1\boldsymbol{v} \,. \tag{A.3}$$

On the other hand, from a strictly mechanical viewpoint, rotation matrices can be interpreted as coordinates of a rotation tensor, the only intrinsic entity representing a rotation [62]. Thus, as a rotation applied to a vector $\boldsymbol{a}$ transforms the vector to a new vector $\boldsymbol{b}$, in the same way the corresponding rotation matrix ${}^0\boldsymbol{R}_1$ multiplied by the coordinates of $\boldsymbol{a}$ gives the coordinates of $\boldsymbol{b}$, where $\boldsymbol{b}$ can be obtained by rotating $\boldsymbol{a}$ about a special vector $\boldsymbol{r}$, which is the eigenvector of the rotation matrix. In fact, $\boldsymbol{a}$ and $\boldsymbol{b}$ become parallel when $\boldsymbol{b} = \boldsymbol{r}$ and $\boldsymbol{a} = \lambda\boldsymbol{r}$, hence the coordinates of $\boldsymbol{r}$ in the base frame are the solution of the eigenproblem

$$ {}^0\boldsymbol{R}_1 {}^0\boldsymbol{r} = \lambda {}^0\boldsymbol{r} \,. \tag{A.4}$$

The solutions of the characteristic equation

$$ \det\left({}^0\boldsymbol{R}_1 - \lambda\boldsymbol{I}\right) = 0 \tag{A.5}$$

are the eigenvalues of the rotation matrix, i.e., $\lambda = 1, \mathrm{e}^{\mathrm{i}\vartheta}, \mathrm{e}^{-\mathrm{i}\vartheta}$ are found, where i is the imaginary unit. To show that the real eigenvector ${}^0\boldsymbol{r}$ corresponding to the real eigenvalue 1 represents the geometric axis about which $\boldsymbol{a}$ must be rotated to match $\boldsymbol{b}$, it must be verified that it has the same coordinates in the frames 0 and 1. This is obvious from the first interpretation of the rotation matrix (A.3), since it is

$$ {}^0\boldsymbol{R}_1 {}^1\boldsymbol{r} = {}^0\boldsymbol{r} \,, \tag{A.6}$$

but, at the same time, since ${}^0\boldsymbol{r}$ is the eigenvector corresponding to $\lambda = 1$, it is

$$ {}^0\boldsymbol{R}_1 {}^0\boldsymbol{r} = {}^0\boldsymbol{r} \,, \tag{A.7}$$

and from the non-singularity of ${}^0\boldsymbol{R}_1$ it follows that

$$^{0}\boldsymbol{r} = {}^{1}\boldsymbol{r} \,. \tag{A.8}$$

This proves the invariance of the real eigenvector of a rotation matrix and thus its geometric meaning in the sense explained before. Also the variable ϑ has a geometric meaning, in fact it is the amount of the rotation about the axis $\boldsymbol{r}$ represented by the rotation matrix ${}^{0}\boldsymbol{R}_1$. The superscript of $\boldsymbol{r}$ can be dropped, since it has the same coordinates in either the body fixed frame and the base frame.

By assuming that the eigenvector $\boldsymbol{r}$ is a unit vector and by exploiting its geometric meaning, it is possible to find a relationship with the rotation matrix, called the *Rodrigues' formula*

$${}^{0}\boldsymbol{R}_1 = \boldsymbol{r}\boldsymbol{r}^{\mathrm{T}} + (\boldsymbol{I} - \boldsymbol{r}\boldsymbol{r}^{\mathrm{T}})\cos\vartheta + \boldsymbol{S}(\boldsymbol{r})\sin\vartheta \,, \tag{A.9}$$

where $\boldsymbol{S}(\cdot)$ is the skew-symmetric matrix performing the vector product, defined as

$$\boldsymbol{S}(\boldsymbol{r}) = \begin{bmatrix} 0 & -r_3 & r_2 \\ r_3 & 0 & -r_1 \\ -r_2 & r_1 & 0 \end{bmatrix} . \tag{A.10}$$

Some useful properties of the skew-symmetric operator are summarized in [55].

A.1.2 Unit quaternion

The term *quaternion* was introduced by Hamilton in 1840, about 70 years after Euler proposed a four-parameter representation of rigid body orientation. From a strictly mathematical viewpoint, quaternions can be seen as a generalization of complex numbers and they allow a global parameterization of the group $SO(3)$; an element of the quaternions' set is defined as

$$\mathcal{Q} = \eta + \epsilon_1 \mathrm{i} + \epsilon_2 \mathrm{j} + \epsilon_3 \mathrm{k}, \quad \epsilon_n \in \mathrm{I\!R}, \quad n = 1,..,3 \,, \tag{A.11}$$

where the symbols $\mathrm{i}, \mathrm{j}, \mathrm{k}$ are the generalized imaginary units. A more compact and useful notation is $\mathcal{Q} = \{\eta, \boldsymbol{\epsilon}\}$, where η is the *scalar part* and $\boldsymbol{\epsilon}$ is the *vector part* of the quaternion. Under the following *quaternion product* operation

$$\mathcal{Q}_1 * \mathcal{Q}_2 = (\eta_1\eta_2 - \boldsymbol{\epsilon}_1^{\mathrm{T}}\boldsymbol{\epsilon}_2, \eta_1\boldsymbol{\epsilon}_2 + \eta_2\boldsymbol{\epsilon}_1 + \boldsymbol{S}(\boldsymbol{\epsilon}_1)\boldsymbol{\epsilon}_2) \tag{A.12}$$

it is possible to show that the set of quaternions is a group. In this group, the identity element is $\mathcal{I} = \{1, \boldsymbol{0}\}$, in fact it is

$$\mathcal{I} * \mathcal{Q} = \mathcal{Q} * \mathcal{I} = \mathcal{Q}, \quad \forall \mathcal{Q} \,. \tag{A.13}$$

Moreover, the inverse of a quaternion is $\mathcal{Q}^{-1} = \{\eta, -\boldsymbol{\epsilon}\}/||\mathcal{Q}||^2$, where

$$||\mathcal{Q}||^2 = \eta^2 + \epsilon_1^2 + \epsilon_2^2 + \epsilon_3^2. \tag{A.14}$$

The subset of all the quaternions with unitary norm is called the group of *unit quaternions*, i.e.

$$\eta^2 + \boldsymbol{\epsilon}^{\mathrm{T}}\boldsymbol{\epsilon} = 1\,. \tag{A.15}$$

In order to relate unit quaternions and rotations, given a rotation matrix (with its real unit eigenvector $\boldsymbol{r}$) it is possible to choose

$$\eta = \cos\frac{\vartheta}{2} \tag{A.16a}$$

$$\boldsymbol{\epsilon} = \boldsymbol{r}\sin\frac{\vartheta}{2}, \tag{A.16b}$$

in fact it is straightforward to verify that this is a unit quaternion. The main feature of this choice is its geometric interpretation, since the vector part of the quaternion is parallel to the eigenvector of the rotation matrix, hence it is parallel to the axis of the rotation represented by the rotation matrix. It is worth pointing out that the definition of such four parameters was made by Euler 70 years before Hamilton would have introduced the notion of quaternion; for this reason unit quaternions are often called *Euler parameters*. By taking into account the definition of Euler parameters (A.16a),(A.16b), the Rodrigues' formula (A.9) can be rewritten as

$${}^{0}\boldsymbol{R}_1 = 2\boldsymbol{\epsilon}_{10}\boldsymbol{\epsilon}_{10}^{\mathrm{T}} + (2\eta_{10}^2 - 1)\boldsymbol{I} + 2\eta_{10}\boldsymbol{S}(\boldsymbol{\epsilon}_{10})\,. \tag{A.17}$$

From (A.17), it is easy to notice that for each rotation matrix, there exist two distinct unit quaternions representing the rotation, in fact quaternions $\{\eta_{10}, \boldsymbol{\epsilon}_{10}\}$ and $\{-\eta_{10}, -\boldsymbol{\epsilon}_{10}\}$ give the same rotation matrix, i.e., the set of unit quaternion is a two-to-one covering of $SO(3)$. But, if the rotation angle ϑ is restricted to the interval $[-\pi, \pi]$, namely $\eta \geq 0$, the covering is a one-to-one. In any case, unit quaternions represent a global parametrization of $SO(3)$, differently from Euler angles, hence it is a singularity-free representation. Notice that, in view of (A.8), the reference frame (the superscript) of the vector part of a unit quaternion can be one of the two frames at the subscripts, and it will be hereafter omitted.

Compared to the Euler angles representation, the unit quaternion can be used exactly in the same way of rotation matrix, because it is possible to define a *quaternion algebra* [26]. In this way, successive rotations or reference frame change are meaningful, as summarized in Table A.1.

Table A.1. Rotation matrix and unit quaternion corresponding operations

Rotation matrix	*Unit quaternion*
${}^{a}\boldsymbol{R}_b$	$\mathcal{Q}_{ba} = \{\eta_{ba}, \boldsymbol{\epsilon}_{ba}\}$
${}^{a}\boldsymbol{v} = {}^{a}\boldsymbol{R}_b\,{}^{b}\boldsymbol{v}$	${}^{a}\boldsymbol{v} = {}^{b}\boldsymbol{v} + 2\eta_{ba}\boldsymbol{S}(\boldsymbol{\epsilon}_{ba})\,{}^{b}\boldsymbol{v} + 2\boldsymbol{S}(\boldsymbol{\epsilon}_{ba})\boldsymbol{S}(\boldsymbol{\epsilon}_{ba})\,{}^{b}\boldsymbol{v}$
${}^{a}\boldsymbol{R}_c = {}^{a}\boldsymbol{R}_b\,{}^{b}\boldsymbol{R}_c$	$\mathcal{Q}_{ca} = \{\eta_{ba}\eta_{cb} - \boldsymbol{\epsilon}_{ba}^{\mathrm{T}}\boldsymbol{\epsilon}_{cb}, \eta_{ba}\boldsymbol{\epsilon}_{cb} + \eta_{cb}\boldsymbol{\epsilon}_{ba} + \boldsymbol{S}(\boldsymbol{\epsilon}_{ba})\boldsymbol{\epsilon}_{cb}\}$
${}^{a}\boldsymbol{R}_b^{-1} = {}^{a}\boldsymbol{R}_b^{\mathrm{T}} = {}^{b}\boldsymbol{R}_a$	$\mathcal{Q}_{ba}^{-1} = \{\eta_{ba}, -\boldsymbol{\epsilon}_{ba}\}$

Algorithm to extract the unit quaternion from a rotation matrix. The quaternion corresponding to a given rotation matrix $\boldsymbol{R} = \{R_{ij},\, i = 1,..,3,\, j = 1,..,3\}$ is [64]

$$\eta = \frac{1}{2}\sqrt{R_{11} + R_{22} + R_{33} + 1} \tag{A.18a}$$

$$\boldsymbol{\epsilon} = \frac{1}{2}\begin{bmatrix} \mathrm{sgn}(R_{32} - R_{23})\sqrt{R_{11} - R_{22} - R_{33} + 1} \\ \mathrm{sgn}(R_{13} - R_{31})\sqrt{R_{22} - R_{33} - R_{11} + 1} \\ \mathrm{sgn}(R_{21} - R_{12})\sqrt{R_{33} - R_{11} - R_{22} + 1} \end{bmatrix}, \tag{A.18b}$$

where the choice $\eta \geq 0$ has been made, and conventionally, $\mathrm{sgn}(x) = -1$ for $x < 0$ and $\mathrm{sgn}(x) = 1$ for $x \geq 0$. It can be recognized that the extraction of the quaternion is always well-defined.

A.2 Mutual Orientation

In every feedback control algorithm the 'difference' between two quantities is usually computed, as it is useful to quantify how much a quantity is different from another. Therefore, the *orientation displacement* between two frames Σ_a and Σ_b with respect to a reference frame Σ_0 has to be defined. Let ${}^0\boldsymbol{R}_a$ and ${}^0\boldsymbol{R}_b$ be the rotation matrices of the two frames.

The first possible choice to quantify the mutual orientation (or the orientation displacement) of the two frames is defined as the rotation matrix describing the relative rotation between the two frames

$${}^a\boldsymbol{R}_b = {}^0\boldsymbol{R}_a^{\mathrm{T}}\,{}^0\boldsymbol{R}_b = {}^a\boldsymbol{R}_0\,{}^0\boldsymbol{R}_b\,. \tag{A.19}$$

According to the (A.12), this mutual orientation can be equivalently expressed in terms of a unit quaternion as

$$\begin{aligned} \mathcal{Q}_{ba} &= \{\eta_{ba}, \boldsymbol{\epsilon}_{ba}\} = \mathcal{Q}_{a0}^{-1} * \mathcal{Q}_{b0} \\ &= \{\eta_{a0}\eta_{b0} + \boldsymbol{\epsilon}_{a0}^{\mathrm{T}}\boldsymbol{\epsilon}_{b0}, \eta_{a0}\boldsymbol{\epsilon}_{b0} - \eta_{b0}\boldsymbol{\epsilon}_{a0} - \boldsymbol{S}(\boldsymbol{\epsilon}_{a0})\boldsymbol{\epsilon}_{b0}\}\,. \end{aligned} \tag{A.20}$$

The second possible choice to compute the mutual orientation between two frames is

$$\boldsymbol{R}_{ba} = {}^0\boldsymbol{R}_b\,{}^0\boldsymbol{R}_a^{\mathrm{T}} = {}^0\boldsymbol{R}_b\,{}^a\boldsymbol{R}_0\,, \tag{A.21}$$

which is still a rotation matrix but it does not describe the relative rotation of the two frames, hence, it has been denoted with two subscripts and no superscript.

Equivalently, the corresponding unit quaternion is given by

$$\mathcal{Q}_{b0} * \mathcal{Q}_{a0}^{-1} = \{\eta_{a0}\eta_{b0} + \boldsymbol{\epsilon}_{a0}^{\mathrm{T}}\boldsymbol{\epsilon}_{b0}, \eta_{a0}\boldsymbol{\epsilon}_{b0} - \eta_{b0}\boldsymbol{\epsilon}_{a0} + \boldsymbol{S}(\boldsymbol{\epsilon}_{a0})\boldsymbol{\epsilon}_{b0}\}\,. \tag{A.22}$$

It is possible to show [26] that the vector part of this quaternion is the vector part of the quaternion in (A.20) expressed in the base frame 0, while in (A.20) it was expressed in frame Σ_a or frame Σ_b.

A.3 Angular Velocity

Let ${}^a\boldsymbol{R}_b(t)$ denote a generic time-varying rotation matrix, if $\boldsymbol{\omega}_{ba}$ is the *angular velocity vector* of the frame Σ_b respect to the frame Σ_a, it can be shown that [54]

$$ {}^a\dot{\boldsymbol{R}}_b = \boldsymbol{S}({}^a\boldsymbol{\omega}_{ba}){}^a\boldsymbol{R}_b = {}^a\boldsymbol{R}_b\boldsymbol{S}({}^b\boldsymbol{\omega}_{ba}) \,, \tag{A.23} $$

where ${}^a\boldsymbol{\omega}_{ba}$ and ${}^b\boldsymbol{\omega}_{ba}$ are two different expressions of the same vector ($\boldsymbol{\omega}_{ba}$), one in the frame Σ_a and the other in the frame Σ_b. If the roles of the two frames are inverted then it can be shown that $\boldsymbol{\omega}_{ba} = -\boldsymbol{\omega}_{ab}$ [62], and thus

$$ {}^a\dot{\boldsymbol{R}}_b = -\boldsymbol{S}({}^a\boldsymbol{\omega}_{ab}){}^a\boldsymbol{R}_b = -{}^a\boldsymbol{R}_b\boldsymbol{S}({}^b\boldsymbol{\omega}_{ab}) \,. \tag{A.24} $$

It is worth noticing that the sequence of subscripts in the angular velocity vector ends with the reference frame. In (A.23) the second subscript refers to the superscript of the rotation matrix, while, in (A.24) it refers to the subscript of the rotation matrix.

Let now $\Sigma_a, \Sigma_b, \Sigma_c$ be three rotating frames. Then, with reference to (A.23), the property of *angular velocity composition* holds [62]

$$ {}^c\boldsymbol{R}_a = {}^c\boldsymbol{R}_b{}^b\boldsymbol{R}_a \quad \Rightarrow \quad \boldsymbol{\omega}_{ac} = \boldsymbol{\omega}_{bc} + \boldsymbol{\omega}_{ab} \,, \tag{A.25} $$

from which it can be derived the relative angular velocity between the two frames Σ_b and Σ_c

$$ {}^c\boldsymbol{R}_b = {}^c\boldsymbol{R}_a{}^a\boldsymbol{R}_b \quad \Rightarrow \quad \boldsymbol{\omega}_{bc} = \boldsymbol{\omega}_{ac} - \boldsymbol{\omega}_{ab} \,. \tag{A.26} $$

With reference to the rigid body orientation represented via a unit quaternion, it can be found the relationship between the angular velocity and the time derivative of the quaternion.

By taking the time derivative of (A.17) and accounting for (A.23), it can be shown that the angular velocity of the body-fixed frame with respect to the base frame is given by [32, 3]

$$ {}^0\boldsymbol{\omega}_{10} = 2\left((\eta_{10}\boldsymbol{I} + \boldsymbol{S}(\boldsymbol{\epsilon}_{10}))\dot{\boldsymbol{\epsilon}}_{10} - \dot{\eta}_{10}\boldsymbol{\epsilon}_{10}\right) . \tag{A.27} $$

From (A.27) and the time derivative of the normalization condition of unit quaternion, it is not hard to find the so called equation of *unit quaternion propagation*

$$ \dot{\eta}_{10} = -\frac{1}{2}\boldsymbol{\epsilon}_{10}^{\mathrm{T}}\,{}^0\boldsymbol{\omega}_{10} \tag{A.28a} $$

$$ \dot{\boldsymbol{\epsilon}}_{10} = \frac{1}{2}\boldsymbol{E}(\eta_{10}, \boldsymbol{\epsilon}_{10}){}^0\boldsymbol{\omega}_{10} \,, \tag{A.28b} $$

with

$$ \boldsymbol{E} = \eta_{10}\boldsymbol{I} - \boldsymbol{S}(\boldsymbol{\epsilon}_{10}) \,. \tag{A.29} $$

By looking at this equation it can be recognized that, given a time-varying rotation of a rigid body, i.e., its angular velocity, the time derivative of unit quaternion representing the rigid body rotation is always well-defined.

It is worth noticing that even though the vector part of the quaternion can be expressed indifferently in the base frame or in the body-fixed frame, the angular velocity is necessarily expressed in the base frame. It's possible to show [62] that, if the angular velocity of the body respect to the base frame is expressed in the body-fixed frame, equation (A.27) becomes

$$ {}^{1}\boldsymbol{\omega}_{10} = 2\left((\eta_{10}\boldsymbol{I} - \boldsymbol{S}(\boldsymbol{\epsilon}_{10}))\dot{\boldsymbol{\epsilon}}_{10} - \dot{\eta}_{10}\boldsymbol{\epsilon}_{10}\right) \tag{A.30} $$

while the propagation equation is written as

$$ \dot{\eta}_{10} = -\frac{1}{2}\boldsymbol{\epsilon}_{10}^{\mathrm{T}}\,{}^{1}\boldsymbol{\omega}_{10} \tag{A.31a} $$

$$ \dot{\boldsymbol{\epsilon}}_{10} = \frac{1}{2}\boldsymbol{E}_1(\eta_{10}, \boldsymbol{\epsilon}_{10})\,{}^{1}\boldsymbol{\omega}_{10} \tag{A.31b} $$

with

$$ \boldsymbol{E}_1 = \eta_{10}\boldsymbol{I} + \boldsymbol{S}(\boldsymbol{\epsilon}_{10}) \,. \tag{A.32} $$

Notice that $\{\eta_{10}, \boldsymbol{\epsilon}_{10}\}$ is the quaternion corresponding to the rotation matrix ${}^{0}\boldsymbol{R}_1$.

If a relative rotation between two generic frames Σ_c and Σ_d is considered, it is necessary to take into account the property of angular velocity composition (A.26)

$$ {}^{d}\boldsymbol{R}_c = {}^{d}\boldsymbol{R}_0{}^{0}\boldsymbol{R}_c \quad \Rightarrow \quad \boldsymbol{\omega}_{dc} = \boldsymbol{\omega}_{d0} - \boldsymbol{\omega}_{c0} \quad \Rightarrow \quad \boldsymbol{\omega}_{cd} = \boldsymbol{\omega}_{c0} - \boldsymbol{\omega}_{d0} \,. \tag{A.33} $$

Then, it can be shown that

$$ {}^{d}\dot{\boldsymbol{R}}_c = \boldsymbol{S}(\Delta^{d}\boldsymbol{\omega}_{cd})\,{}^{d}\boldsymbol{R}_c, \quad \text{with} \quad \Delta^{d}\boldsymbol{\omega}_{cd} = {}^{d}\boldsymbol{\omega}_{c0} - {}^{d}\boldsymbol{\omega}_{d0} \,, \tag{A.34} $$

thus, the $\Delta^{d}\boldsymbol{\omega}_{cd} = {}^{d}\boldsymbol{\omega}_{cd}$ is the relative angular velocity between the frame Σ_c and the frame Σ_d. This means that to compute the 'difference' between two orientations, the mutual orientation has to be used (via rotation matrix composition or quaternion product), while at velocity level the simple algebraic difference of angular velocities can be adopted. For the purpose of precision, it can be stressed that if $\{\eta_{cd}, \boldsymbol{\epsilon}_{cd}\}$ is the quaternion extracted from the rotation matrix ${}^{d}\boldsymbol{R}_c$, the correct quaternion propagation that relates the time derivative of such a quaternion with the relative angular velocity is given by the (A.28a),(A.28b), in which $\boldsymbol{\epsilon}_{cd}$ and ${}^{d}\boldsymbol{\omega}_{cd}$ must be used.

B. Real-Time Implementation Notes

This appendix contains some notes about the real-time implementation on the COMAU C3G-9000 (open version) control unit of the interaction control algorithms described in this book. Even though some specific details can depend on the specific architecture, most of the concepts are general and of crucial importance when a novel, and thus "not well-assessed", control algorithm has to be implemented on an industrial robot.

B.1 The main control loop

The C3G-9000 open version [31] is a PC-based open control architecture, in the sense that a standard PC is "linked" to the industrial controller through a bus-to-bus adapter and it can exchange information with the industrial control unit according to several operating modes, as already explained in Chapter 1. Only the more "open" operating mode will be considered here, i.e. when the PC has direct access to the robot motors' driving unit and it can specify a desired current for each motor every millisecond.

B.1.1 C3G and PC interfacing

When a single system is controlled by two different units, a coordination mechanism has to be conceived. In the available architecture, the C3G unit and the PC can exchange information by means of a shared memory where *every millisecond* (which is therefore, the sampling time of the digital control loop) the C3G executes the following operations

- set the flag `IntActive`
- write the values of the motor shaft angular positions (in DAC units)
- read the desired values for the motor current set-points (in DAC units)

and the PC has to

- reset the flag `IntActive`
- read the values of the motor shaft angular positions (in DAC units)
- compute the current set-points for the next time interval
- write the desired values for the motor current set-points (in DAC units)

- set the watchdog flag

To fulfill the real-time constraint, the PC must execute the above steps within 700 μs. Otherwise, due to the watchdog mechanism, the system reaches an alarm state and the robot drives are switched off and brakes are activated. The watchdog timer, as in every real-time system, is aimed at preventing harmful consequences of unpredictable events, which can occur during the normal operation, e.g. when the PC crashes, and thus is no longer able to write a correct current set-point to be actuated. The sequence outlined above is depicted in Fig. B.1.

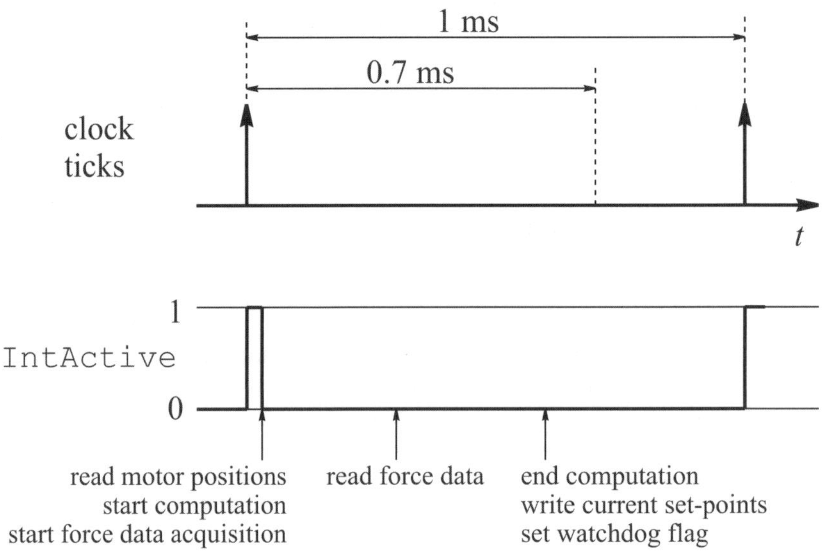

Fig. B.1. Timing diagram of the real-time control loop

A particular discussion is in order when the robotic task requires the cooperation of both robots and the cooperative control strategies proposed in Chapter 5 have to be implemented. The main problem is the synchronization of the two industrial control units with the PC, in fact they have two separate clocks. Since the two clock frequencies are slightly different a time drift between the two trains of clock ticks is experienced. This requires a preliminary phase aimed at waiting until the two clocks are almost aligned, then the control loop can start. Of course, such a solution can guarantee a limited time interval useful for the experiment, in fact the experiment can continue until the time difference between the two clocks is less than 500 μs, otherwise one of the watchdog timers would be triggered. Notice how this solution imposes a minimum sampling time for the control algorithm of 2 ms (twice the time between two clock ticks from the same control unit). A more detailed description of the synchronization issue in cooperative setups can be found in [21].

B.1.2 Force sensor interfacing

The simple operations listed above suffice only if a motion control loop has to be implemented. But, if an interaction control algorithm is being considered, the necessity of interfacing with a force sensor comes up. The ATI force/torque sensor available at the PRISMA Lab can provide the user with a full six components force measurement every millisecond, but special care has to be adopted to optimize the time usage.

In detail, the force sensor is interfaced via a parallel link to the ISA bus of the PC. To obtain the force measurement, the PC has to send a data acquisition request, which starts the data A/D conversion on the sensor conditioning electronics; after a time lapse of about 250 μs, the six components of the force are available in a memory buffer. Since the conversion is executed on remote hardware, during the conversion time interval, the PC can continue its elaboration; hence, to avoid simply waiting the end of conversion, it is convenient to ask for the start of conversion at the beginning of the control cycle, perform all the computations not requiring force measurements, e.g. kinematics, dynamic model compensation, then read the force data and, finally, complete the control algorithm. The timing of these operations is outlined in Fig. B.1.

B.1.3 Safety checks

Even though the use of a watchdog timer is a good solution for a safe operation of a real-time system, it cannot be considered enough to prevent every dangerous situation. In every research lab, where a lot of effort (and money!) has been spent to setup the experimental test bed, the first rule is to avoid any type of damage to the instrumentation (and, possibly, to the researchers...). Therefore, the *safety* issue is as important as the performance of the new control algorithm to be tested. In the C3G open control architecture a lot of safety checks can be carried out to monitor the system functioning and to prevent damage. The following safety checks are available within the C3G open control architecture:

- set and check joint limits
- set and check maximum joint velocities
- set and check maximum instantaneous currents
- set and check maximum sustained currents
- set and check maximum forces and torques

The first type of safety check is particularly useful when an interaction task is tested, in fact limiting the robot workspace allows to prevent damage to the environment, to the robot and especially to the force sensor. Of course, monitoring also contact forces and moments allows a safer contact phase. Typically, interaction tasks evolve with limited velocities, hence monitoring joint velocities is also useful to detect undesirable situations during the task execution.

For the purpose of preventing damage of robot motors and motor drives, it is opportune to monitor the motor currents, whose instantaneous values should not exceed maximum fixed levels. Sometimes, it is also useful to limit the maximum sustained value of the currents, to prevent overheating of robot servo motors and their driving electronics.

Two different decisions can be taken in case one of the above checks fails, namely immediately stop the robot by invoking a special emergency routine, or set an error flag to exit from the main control loop and switch off the drives (see also the C-code listing in the next section). The former approach is faster and thus safer, but requires a complete reboot of the system. The latter approach implies a delay of one sampling period, but leads the system to a less critical state, which does not require a complete reboot. The best policy is probably to decide on the particular safety check which failed, e.g., if one of the motor currents exceeded the maximum sustained value, an immediate stop is not necessary, on the contrary if the force is over the maximum allowed value, an immediate stop is advisable.

B.2 Writing the C-code

To clarify all the details described above, it is useful to analyse a sample C-code listing for a single robot experiment.

After the initialization of the force sensor and of the communication with the C3G control unit, it is necessary to allocate the memory buffers where the measured data will be saved during the execution of the experiment. Of course, it is not possible to record these data directly on disk, since every I/O operation takes a long time, thus, during a real-time operation no I/O function is allowed inside the main control loop and only accesses to RAM memory should be made.

After the drives have been switched on, the first cycle is devoted to control algorithm initialization, then, before starting the main control loop, the watchdog timer can be enabled, and then the watchdog flag must be set every millisecond.

As explained in the previous section, it is advisable to immediately ask for force data acquisition; on the other hand, once the `IntActive` flag is high, motor position measurements are already available, so they can be read immediately, and the PC can start computation of the part of the control algorithm not requiring force information.

Before starting execution of the planned motion/force trajectory, it is necessary to keep the initial position for a while, to let the mechanical joint brakes be completely released.

Based on the motor shaft positions and on the desired trajectory, it is possible to start control algorithm computation, but it can be completed only after force measurement acquisition. The reading of six components of contact force and moment is performed inside the routine `CalcCurrSP`. Notice that also the safety routines are called inside the `CalcCurrSP` function, which returns an error flag.

Once the current set-point have been computed, they can be written in the shared memory, and they will be actuated within the sampling interval. The control loop proceeds with storing of measured data in the allocated memory buffers and ends with setting of the watchdog flag.

The main control loop can be broken if one of the following conditions is verified: the counter exceeds the maximum value fixed for the experiment, the user press a key, an error condition is detected.

When the loop ends, before switching off the drives, which takes more than one millisecond, the watchdog timer must be disabled. Finally, the communication with the C3G unit is terminated and the content of all the memory buffers is recorded on files. An error detection routine is devoted to decode the error flag `Err` and provides the user with a message.

```
void main(void) {

   /* Declarations/definitions of all the necessary variables */
   [omitted]

   /* Force/Torque sensor initialization */
   InitStatus = FTInit();

   /* C3G Initialization */
   SetRobotType(SMART_3_S);
   SetJointLimits(Qmin,Qmax);
   B3Init();

   /* Initialize memory buffers for data recording */
   InitRecordMotorPositions(NumRecMotPos);
   InitRecordCurrents(NumRecCurrSP);
   FTInitRecordData(nFTSamples);

   /* DRIVE ON */
   DriveOn_OK = (DriveOn(TIME_OUT) == 0);

   if(DriveOn_OK) {
       /* Initialize control algorithm */
       IntActive = 0;        /* reset interrupt flag */
       while(IntActive == 0)/* synchronize with interrupt signal */
         ;
       ReadMotorPositions(MotorPositions);
       InitCalcCurrSP(MotorPositions,q0,x0,J0);

       /* Enable watchdog timer */
       EnableWatchdog();
```

```
    /* Set watchdog flag */
    SetWatchdog();

    /* MAIN CONTROL LOOP */
    IntActive = 0;
    do {
      if(IntActive) {
         IntActive = 0;

    /* Ask for force sensor data */
    AskStatus = FTAskData();

    /* Read motor shaft positions from shared memory */
    ReadMotorPositions(MotorPositions);

    /* Initialize desired force/motion trajectory */
    InitTrajh(hd_c,timef,counter);
    InitTrajQu(xd,xpd,time,counter);

    /* Keep initial position until brakes are released */
    Pause1(&time,Tass);

    /* Generate desired motion trajectory */
    GenerateSegmQuatVG(lx1,ly1,lz1,thd1,rx1,ry1,rz1,
                       &time,dur1,xd,&th1);
    Derivate(xd,xpd,xppd,th1,rx,ry,rz);

    /* Generate desired force trajectory */
    Pause1(&timef,2000.0);
    GenerateHd(fxd,fyd,fzd,mxd,myd,mzd,&timef,2500.0,hd_c);

    /* Compute current set-points */
    Err=CalcCurrSP(MotorPositions,q0,x0,J0,xd,xpd,xppd,hd_c,
                   Currents,AskStatus);

    /* Send current set-points to shared memory */
    WriteCurrents(Currents);

    /* Save measured data in memory buffers */
    FTSaveData();
    SaveMotorPositions(MotorPositions);
    SaveCurrents(Currents);
    counter++;

    /* Set the watchdog flag */
    SetWatchdog();
   }
  } while((counter<MaxCount) && (!kbhit()) && (!Err));
  /* END OF MAIN CONTROL LOOP */

  /* Disable watchdog timer */
  DisableWatchdog();
}
```

```
    /* DRIVE OFF */
    DriveOff(TIME_OUT);

    /* Terminate communication with C3G unit */
    B3Terminate();

    /* Record data on disk */
    RecordMotorPositions(namepos);
    RecordCurrents(namecurr);
    FTRecordData(nameft);

    /* Display any error message */
    ErrorDetection(Err);
}
```

The last issue of this appendix consists of a few notes about code optimization. In a real-time application, speed up code execution is the first rule, even if this means duplicating code. In fact, it is well known that modularity is an important feature of software, but jumping to a subroutine takes time. Thus, when the time constraint is very demanding, it can be necessary to renounce the modularity and repeat the code as many times as necessary. A same consideration can be made in respect to loops. Even if a series of repeating instructions can be coded with a loop, by repeating the code of the instruction saves a lot of time, so *loop unrolling* can be necessary, and it is certainly convenient when a processor using instruction-level parallelism with pipelines is adopted. For a deep discussion about code optimization, see e.g. [38] and [2] for Pentium processor programming.

Index

Springer Tracts in Advanced Robotics

Edited by B. Siciliano, O. Khatib, and F. Groen
Published Titles:

Vol. 1: Caccavale, F.; Villani, L. (Eds.)
Fault Diagnosis and Fault Tolerance for Mechatronic Systems: Recent Advances
191 pp. 2002 [3-540-44159-X]

Vol. 2: Antonelli, G.
Underwater Robots: Motion and Force Control of Vehicle-Manipulator Systems
209 pp. 2003 [3-540-00054-2]

Vol. 4: Bicchi, A.; Christensen, H.I.; Prattichizzo, D. (Eds.)
Control Problems in Robotics
296 pp. 2003 [3-540-00251-0]

Printing (Computer to Plate): Saladruck Berlin
Binding: Stürtz AG, Würzburg